Despeckle Filtering for Ultrasound Imaging and Video

Volume II

Selected Applications

Second Edition

Synthesis Lectures on Algorithms and Software in Engineering

Editor
Andreas Spanias, *Arizona State University*

Despeckle Filtering for Ultrasound Imaging and Video, Volume II: Selected Applications, Second Edition
Christos P. Loizou and Constantinos S. Pattichis
2015

Despeckle Filtering for Ultrasound Imaging and Video, Volume I: Algorithms and Software, Second Edition
Christos P. Loizou and Constantinos S. Pattichis
2015

Latency and Distortion of Electromagnetic Trackers for Augmented Reality Systems
Henry Himberg and Yuichi Motai
2014

Bandwidth Extension of Speech Using Perceptual Criteria
Visar Berisha, Steven Sandoval, and Julie Liss
2013

Control Grid Motion Estimation for Efficient Application of Optical Flow
Christine M. Zwart and David H. Frakes
2013

Sparse Representations for Radar with MATLAB ™ Examples
Peter Knee
2012

Analysis of the MPEG-1 Layer III (MP3) Algorithm Using MATLAB
Jayaraman J. Thiagarajan and Andreas Spanias
2011

Theory and Applications of Gaussian Quadrature Methods
Narayan Kovvali
2011

Algorithms and Software for Predictive and Perceptual Modeling of Speech
Venkatraman Atti
2011

Adaptive High-Resolution Sensor Waveform Design for Tracking
Ioannis Kyriakides, Darryl Morrell, and Antonia Papandreou-Suppappola
2010

MATLAB™ Software for the Code Excited Linear Prediction Algorithm: The Federal Standard-1016
Karthikeyan N. Ramamurthy and Andreas S. Spanias
2010

OFDM Systems for Wireless Communications
Adarsh B. Narasimhamurthy, Mahesh K. Banavar, and Cihan Tepedelenliouglu
2010

Advances in Modern Blind Signal Separation Algorithms: Theory and Applications
Kostas Kokkinakis and Philipos C. Loizou
2010

Advances in Waveform-Agile Sensing for Tracking
Sandeep Prasad Sira, Antonia Papandreou-Suppappola, and Darryl Morrell
2008

Despeckle Filtering Algorithms and Software for Ultrasound Imaging
Christos P. Loizou and Constantinos S. Pattichis
2008

Despeckle Filtering for Ultrasound Imaging and Video, Volume II: Selected Applications, Second Edition
Christos P. Loizou and Constantinos S. Pattichis

ISBN: 978-3-031-00396-7   paperback
ISBN: 978-3-031-01524-3   ebook

DOI 10.1007/978-3-031-01524-3

A Publication in the Springer series
SYNTHESIS LECTURES ON ALGORITHMS AND SOFTWARE IN ENGINEERING

Lecture #15
Series Editor: Andreas Spanias, *Arizona State University*
Series ISSN
Print 1938-1727   Electronic 1938-1735

Despeckle Filtering for Ultrasound Imaging and Video

Volume II

Selected Applications

Second Edition

Christos P. Loizou
School of Sciences and Engineering, Intercollege, Cyprus

Constantinos S. Pattichis
University of Cyprus

SYNTHESIS LECTURES ON ALGORITHMS AND SOFTWARE IN ENGINEERING #15

ABSTRACT

In ultrasound imaging and video visual perception is hindered by speckle multiplicative noise that degrades the quality. Noise reduction is therefore essential for improving the visual observation quality or as a pre-processing step for further automated analysis, such as image/video segmentation, texture analysis and encoding in ultrasound imaging and video. The goal of the first book (book 1 of 2 books) was to introduce the problem of speckle in ultrasound image and video as well as the theoretical background, algorithmic steps, and the MATLAB™ code for the following group of despeckle filters: linear despeckle filtering, non-linear despeckle filtering, diffusion despeckle filtering, and wavelet despeckle filtering. The goal of this book (book 2 of 2 books) is to demonstrate the use of a comparative evaluation framework based on these despeckle filters (introduced on book 1) on cardiovascular ultrasound image and video processing and analysis. More specifically, the despeckle filtering evaluation framework is based on texture analysis, image quality evaluation metrics, and visual evaluation by experts. This framework is applied in cardiovascular ultrasound image/video processing on the tasks of segmentation and structural measurements, texture analysis for differentiating between two classes (i.e. normal vs disease) and for efficient encoding for mobile applications. It is shown that despeckle noise reduction improved segmentation and measurement (of tissue structure investigated), increased the texture feature distance between normal and abnormal tissue, improved image/video quality evaluation and perception and produced significantly lower bitrates in video encoding. Furthermore, in order to facilitate further applications we have developed in MATLAB™ two different toolboxes that integrate image (IDF) and video (VDF) despeckle filtering, texture analysis, and image and video quality evaluation metrics. The code for these toolsets is open source and these are available to download complementary to the two monographs.

KEYWORDS

speckle, despeckle, noise filtering, ultrasound, ultrasound imaging, ultrasound video, cardiovascular imaging and video, texture, image and video quality, video encoding, mobile health, carotid artery

To my Family, and
in memory of my father Panayiotis and my mother Eleni

Christos P. Loizou

To my mother, and
in memory of my father Stephanos, and my sister Revecka

Constantinos S. Pattichis

"This is the way we should see Christ. He is our friend, our brother; He is whatever is good and beautiful. He is everything. Yet, He is still a friend and He shouts it out, "You're my friends, don't you understand that? We're brothers. I'm not...I don't hold hell in my hands. I am not threatening you. I love you. I want you to enjoy life together with me."

"Love Christ and put nothing before His Love. He is joy, He is life, He is light. Christ is Everything. He is the ultimate desire, He is everything. Everything beautiful is in Christ."

"The life of the parents is the only thing that makes good children. Parents should be very patient and 'saintlike' to their children. They should truly love their children. And the children will share this love! For the bad attitude of the children, says father Porphyrios, the ones who are usually responsible for it are their parents themselves. The parents don't help their children by lecturing them and repeating to them 'advices', or by making them obeying strict rules in order to impose discipline. If the parents do not become 'saints' and truly love their children and if they don't struggle for it, then they make a huge mistake. With their wrong and/or negative attitude the parents convey to their children their negative feelings. Then their children become reactive and insecure not only to their home, but to the society as well."

Saint Porphyrios (Bairaktaris) the Kapsokalyvite

Source: Wikipedia

Contents

Preface

Speckle is a multiplicative noise that degrades image quality and the visual evaluation in ultrasound and SAR imaging. This necessitates the need for robust despeckling techniques in a wide spectrum of the aforementioned imaging applications. Despeckle filtering applications has been a rapidly emerging research area in recent years. The goal of the first book (book 1 of 2 books) was to introduce the problem of speckle in ultrasound image and video as well as the theoretical background, algorithmic steps, and the Matalb™code for the following group of despeckle filters: linear despeckle filtering, non-linear despeckle filtering, diffusion despeckle filtering, and wavelet despeckle filtering. The goal of this book (book 2 of 2 books) is to demonstrate the use of a comparative evaluation framework of these despeckle filters (introduced on book 1) on cardiovascular ultrasound image and video processing. More specifically, the despeckle filtering evaluation framework is based on texture analysis, image quality evaluation metrics, and visual evaluation by experts. The filters covered represent only a snapshot of the vast number of despeckle filters and applications published in the literature. The source code of the algorithms presented in this book has been made available on the web, thus enabling researchers to more easily exploit the application of despeckle filtering in their problems under investigation.

The book is organized in five chapters. In Chapter 1 an introduction and review of different despeckle filtering techniques for ultrasound imaging and video is presented, a despeckle filtering evaluation protocol is proposed and selected applications for ultrasound image and video despeckle filtering techniques are outlined. In Chapter 2 we present the application and results of the segmentation of the intima-media complex (IMC), the media-layer (ML) and the intima layer (IL) of the common carotid artery as well as the segmentation of the atherosclerotic carotid plaque from ultrasound images and videos following despeckle filtering. In Chapter 3, we present the results on image and video texture analysis. We provide results from the texture analysis of the IMC and the atherosclerotic carotid plaque performed on a large number of ultrasound images and videos. In Chapter 4 we present results on ultrasound wireless video encoding and transmission which is performed before and after despeckle filtering. Chapter 5 discusses, compares and evaluates the proposed despeckle filtering techniques for image and video and provides an outline of future directions. Finally, at the end of this book, appendices provide details about the IDF and VDF despeckle filtering MATLAB™ toolboxes.

Furthermore, it is noted that for those practicing engineers/scientists whose principal need is to use existing image despeckle filtering technologies and apply them on different type of images or video, there is no simple answer regarding which specific filtering algorithm should be selected without a significant understanding of both the filtering fundamentals, and the application environment under investigation. A number of issues would need to be addressed. These include availability of the image/video to be processed/analyzed, the required level of filtering, the application scope (general-purpose or application-specific), the application goal (for extracting features from the image or for visual enhancement), the allowable computational complexity, the allowable implementation complexity, and the computational requirements (e.g., real-time or offline). We believe that a good understanding of the contents of this book can help the readers make the right choice of selecting the most appropriate filter for the application under development. Furthermore, the despeckle filtering evaluation protocol documented in Table 1.4 could also be exploited.

This book is intended for all those working in the field of image and video processing technologies, and more specifically in medical imaging and in ultrasound image and video preprocessing and analysis. It provides different levels of material to researchers, biomedical engineers, computing engineers, and medical imaging engineers interested in developing imaging systems with better quality images, limiting the corruption of speckle noise.

We wish to thank all the members of our carotid ultrasound imaging team, for the long discussions, advice, encouragement, and constructive criticism they provided to us during the course of this research work. First of all we would like to express our sincere thanks to Emeritus Prof Andrew Nicolaides, of the Faculty of Medicine, Imperial College of Science, Technology and Medicine, UK, and founder of the Vascular Screening and Diagnostic Centre in Cyprus. Furthermore, we would like to express our thanks to Dr Marios Pantziaris, consultant neurologist, at the Cyprus Institute of Neurology and Genetics, Dr Theodosis Tyllis, consultant physician in the private sector in Cyprus, Associate Professor Efthyvoulos Kyriakou, at the Frederick University, Cyprus, Dr Christodoulos Christodoulou, Research Associate at the University of Cyprus, and Associate Professor Marios Pattichis, University of New Mexico, USA. Last but not least, we would like to thank, Prof Andreas Spanias, Arizona State University, USA, for his proposal and encouragement in writing this book, and to Joel Claypool, and the rest of the staff at Claypool publishing house, for their understanding, patience and support in materializing this project.

This work was partly funded through the projects *Integrated System for the Support of the Diagnosis for the Risk of Stroke (IASIS, 2002-2005)*, and *Integrated System for the Evaluation of Ultrasound Imaging of the Carotid Artery (TALOS, 2003-2005)*, funded by the Research Promotion Foundation of Cyprus. Furthermore, partial funding and support was also obtained from both the Cardiovascular Disease Educational and Research Trust (CDER Trust), UK, and the CDER Trust, Cyprus.

We hope that this book will be a useful reference for all the readers in this important field of research and to contribute to the development and implementation of innovative imaging and video systems enabling the provision of better quality images.

Christos P. Loizou and Constantinos S. Pattichis
August 2015

List of Symbols

C	Speckle Index
σ	Standard deviation
σ^2	Variance
σ^3	Skewness
σ^4	Kurtosis
σ_n	Standard deviation of the noise
ρ	Correlation Coefficient
β	Shape parameter
IMT_{mean}	Mean value of the IMT
$IMT_{\min}$	IMT minimum value
$IMT_{\max}$	IMT maximum value
IMT_{median}	IMT median value
Q	Mathematically defined universal quality index
$feat\_dis_i$	Percentage distance
$Score\_Dis$	Score distance between two classes (asymptomatic, symptomatic)
dis_{zc}	Distance between asymptomatic and symptomatic images
m_{i1}, m_{i2}	Mean values of two classes (asymptomatic, symptomatic)
$\sum Var$	Sum variance
$\sum Entr$	Sum Entropy
p	Significance level for a statistical test
Hz, KHz, MHz	Hertz, Kilohertz, Megahertz
$cm/s, cm/s$	Metres per second, centimetres per second
μ	Mean
N	Number of scatterers within a resolution cell
N_{feat}	Number of features in the feature set
$s_e = \sigma_{IMT}/\sqrt{2}$	Inter-observer error
σ_{IMT}	IMT standard deviation

List of Abbreviations

IDF	Image despeckle filtering toolbox
VDF	Video despeckle filtering toolbox
IVUS	Intra Vascular ultrasound
OCT	Optical coherence tomography
SAR	Synthetic aperture radar
FFT	Fast Fourier Transform
DsFlsmv	Mean and variance local statistics despeckle filter
DsFlsmvsk1d	Mean, variance, skewness, kurtosis 1D local statistics despeckle filter
DsFlsmvsk2d	Mean variance, higher moments local statistics despeckle filter
DsFlsminsc	Minimum speckle index homogeneous mask despeckle filter
DsFwiener	Wiener despeckle filter
DsFmedian	Median despeckle filter
DsFls	Linear scaling of the gray level values despeckle filter
DsFca	Linear scaling of the gray-levels despeckle filter
DsFlecasort	Linear scaling and sorting despeckle filter
DsFhomog	Most homogeneous neighbourhood despeckle filter
DsFgf4d	Geometric despeckle filter
DsFhomo	Homomorphic despeckle filter
DsFhmedian	Hybrid median filter
DsFkuwahara	Kuwahara despeckle filter
DsFgfminmax	Geometric despeckle filter utilising minimum maximum values
DsFnlocal	Nonlocal filter
DsFad	Perona and Malik anisotropic diffusion filter
DsFsrad	Speckle reducing anisotropic diffusion filter
DsFlsmedcd	Lee diffusion despeckle filter
DsFlsrad	Speckle reducing anisotropic diffusion
DsFnldif	Non-linear anisotropic diffusion despeckle filter
DsFncdif	Non-linear complex diffusion filtering
DsFwaveltc	Wavelet despeckle filter
CCA	Common carotid artery
IMC	Intima media complex
SRAD	Speckle reducing anisotropic diffusion
1D	One-dimensional

2D	Two-dimensional
3D	Three-dimensional
ROI	Area of interest
GUI	Graphical user interface
IMT	Intima-media thickness
IMC	Intima-media complex
IL	Intima layer
ML	Media Layer
ILT	Intima layer thickness
MLT	Media Layer thickness
NPS	No pre-processing
DS	Despeckled
NDS	Normalised despeckled
sd	Standard deviation
GVF	Gradient vector flow
CDC	Carotid diameter during contraction
CDD	Carotid diameter during distension
%CWD	Percentage of carotid wall distension
MMSE	Minimum mean-square error
MSE	Mean square error
MAE	Mean absolute error
MARE	Mean relative absolute error
RMSE	Root mean square error
kNN	The statistical k-nearest-neighbour classifier
PSNR	Peak signal-to-noise radio
SNR	Signal-to-noise radio
COV	Coefficient of Variation
ATL HDI-3000	ATL 3000 ultrasound scanner
ATL HDI-5000	ATL 5000 ultrasound scanner
SSIM	Structural similarity index
VI	Visual inspection
CNR	Contrast to-noise ratio
CPU	Central processing unit
MSE	Mean square error
ENL	Effective number of looks
ρ	Correlation coefficient
FOM	Figure of merit
OSRAD	Oriented speckle reducing anisotropic diffusion
F	FORTRAN

ROI	Area of interest
CCA	Common carotid artery images
IQR	Inter quartile range
M	Manual
MN	Manual normalised
NF	No-filtering
N	Normalised
CV%	Coefficient of variation
se	Intra-observer error
HD	Hausdorff distance
CDC	Carotid diameter during contraction
CDD	Carotid diameter during distension
%CWD	Percentage of carotid wall distension
CIMA	Carotid intima-media area
UIR	User interaction required
MAD	Mean absolute distance
BA	Bland-Altman plots
GVF	Gradient vector flow
FNF	False negative fraction
FPF	False positive fraction
TNF	True negative fraction
TPF	True positive fraction
KI	Williams or kappa index
R	Sensitivity
Sp	Specificity
P	Precision
O	Overlap
F=1-E	Effectiveness measure
AW	Adventitia-wall
PW	Plaque-wall
MAW	Media-Adventitia-wall
NS	Not significantly different
S	Significantly different
RSW	Radial strain at wall
LS	Longitudinal stain at wall
SSW	Shear stain at wall
SSP	Shear strain at plaque
RSP	Radial strain at plaque
AIC	Automatic initial contour

MPD	Mean point distance
GAE	Geometric average error
GF	Geometric filtering
GGVF	Generalised gradient vector flow
GLDS	Gray level difference statistics
HF	Maximum homogeneity
HM	Homomorphic
HVS	Human visual system
ICA	Internal carotid artery
IDM	Inverse difference moment
IDV	Intensity difference vector
LS	Linear scaling
MRI	Magnetic resonance imaging
PET	Position emission tomography
NGTDM	Neighbourhood gray tone difference matrix
ROC	Receiver operating characteristic
SFM	Statistical feature matrix
SGLDM	Spatial gray level dependence matrices
SGLDMm	Spatial gray level dependence matrix mean values
SGLDMr	Spatial gray level dependence matrix range of values
TEM	Laws texture energy measures
FDTA	Fractal dimension texture analysis
FPS	Fourier power spectrum
SOSV	Sum of square variance
SAV	Sum average
ASM	Angular second moment
Err_3, Err_4	Minkwoski metrics
Q	Quality index
SSIN	Structural similarity index
AD	Average difference
SC	Structural content
MD	Maximum difference
kNN	k-nearest neighbour
JPEG	Joint Photographic Experts Group
VSNR	Visual signal-to-noise ratio
IFC	Information fidelity criterion
NQM	Noise quality measure
WSNR	weighted signal-to-noise ratio
HVEC	High efficiency video coding

ITU-T	International telecommunication union-telecommunication sector
VCEG	Video quality experts group
JVT	Joint video team
VCL	Video quality layer
MPEG	Motion picture experts group
NAL	Network abstract layer
FMO	Flexible macroblock ordering
JCT-VC	Joint collaborative team on video coding
SAO	Sample adaptive offset
AMVP	Advanced motion vector prediction
CABAC	Context adaptive binary arithmetic coding
CTU	Coding tree unit
GSM	Global system for mobile communication
1G	1st Generation
2G	2nd Generation
GPRS	General packet radio service
UMTS	Universal mobile telecommunications system
HSDPA	High speed download packet access
HSUPA	High speed uplink packet access
HSPA	High speed packet access
LTE	Long term evolution
ITU-R	International telecommunication union-radio communication sector
QoS	Quality of service
WiMax	Worldwide interoperability for Microwave access
LOS	Line-of-sight
NLOS	Non line-of-sight
HARQ	Hybrid automatic repeat request
OFDM	Orthogonal frequency division multiplexing
SOFDMA	Scalable orthogonal frequency division multiple access
DL	Downlink
UL	Uplink
MIMO	Multiple input multiple output
MAC	Medium access control
PHY	Physical
SC-FDMA	Single carrier frequency division multiple access
FDD	Frequency division duplex
TDD	Time division duplex
ARQ	Automatic repeat request
CoMP	Coordinated multipoint transmission and reception

c-VQA	Clinical video quality assessment
UEP	Unequal error protection
MOS	Mean opinion scores

CHAPTER 1

Introduction and Review of Despeckle Filtering

This chapter provides an introduction and brief overview of selected despeckle filtering techniques for ultrasound imaging and video already presented in Volume I of this book [1]. A despeckle filtering evaluation protocol is proposed, a brief literature review, as well as the image despeckle filtering toolbox (IDF) [2] and the video despeckle filtering (VDF) [3] software toolbox are presented. Moreover, selected applications for ultrasound image and video despeckle filtering techniques are illustrated. The chapter ends with a guide to the book's contents.

1.1   AN OVERVIEW OF DESPECKLE FILTERING TECHNIQUES

In recent years significant technological advancements and progress in image and video processing in a number of areas have been achieved, however, still a number of factors in the visual quality of images, hinder the automated analysis [4], and disease evaluation [5]. These include imperfections of image acquisition instrumentations, natural phenomena, transmission errors, and coding artifacts, which all degrade the quality of image in the form of induced noise [6–14]. Ultrasound imaging and video is a powerful non-invasive diagnostic tool in medicine, but it is degraded by a form of multiplicative noise (speckle), which makes visual observation difficult [1–3, 15–27]. Speckle is mainly found in echogenic areas of the image in the form of a granular appearance that affects texture of the image [28–32], which may carry important information about the shape of tissues and organs. Texture [5, 8, 33] and morphology [34] may provide additional quantitative information of the area under investigation, which may complement the human evaluation and provide additional diagnostic details. It is therefore of interest for the image and video processing community to investigate and apply new image despeckle filtering techniques that can increase the visual perception evaluation and further automate image and video analysis, thus improving the final result. These techniques are usually incorporated into integrated software for image processing applications.

Despeckle filtering has been a rapidly emerging research area in recent years and a significant number of representative studies have been published in numerous journals. The basic principles, the theoretical background and the algorithmic steps of a representative set of despeckle filters were covered in the first volume of this book [1]. In addition, selected representative applications of image and video despeckling covering a variety of ultrasound image and video processing tasks will be presented in the present book.

Table 1.1: An overview of different imaging modalities where despeckle image or video filtering is applied

IMAGING MODALITY	TYPE OF NOISE
Ultrasound	Speckle, Gaussian
Intravascular Ultrasound (IVUS) Imaging	Speckle, Gaussian
Optical Coherence tomography (OCT)	Speckle, Gaussian, white noise
Optical Microscopy	Speckle, shot, dark, red
Synthetic Aperture Radar (SAR) imaging	Speckle, Gaussian, thermal, electronic

An overview of different imaging modalities where despeckle image and video filtering can be applied is presented in Table 1.1, as also documented in the first volume of this book [1]. It should be noted however, that it is not always desirable to remove speckle noise from the images or videos as it can be considered as a natural tissue effect which may provide additional information, especially in the processing and analysis of strain imaging and speckle tracking [35], and in studies of ultrasound tissue characterization [36].

Table 1.2 summarizes the despeckle filtering techniques for ultrasound imaging and video that were presented in detail in the companion volume (Volume I) [1]. These are grouped under the following categories: linear filtering, nonlinear filtering, diffusion filtering and wavelet filtering. Furthermore, in Table 1.2 the main characteristics, the references of the main investigators, and the corresponding filter names for each filtering method are given.

Table 1.3 illustrates summary findings of despeckle filtering in an artificial carotid image (A), a phantom ultrasound image (P) and real ultrasound image of the carotid artery (C), already presented in the companion volume (Volume I) [1]. As given in Table 1.3, filters *DsFlsmv* and *DsFhmedian* improved the statistical and texture features analysis, the measurements and shape features, the image quality evaluation and the optical perception evaluation. This was observed for both filters on the artificial image, the phantom image and the real ultrasound image. In addition, filter *DsFlsminsc* improved the statistical and texture analysis and the image quality evaluation in real carotid ultrasound images. The filter *DsFgf4d* is suitable for improving the visual evaluation in real ultrasound images. The *DsFhomo* showed a small marginal improvement for the statistical image features. Filters *DsFnldif* and *DsFwaveltc* showed similar performance by improving the measurements and shape features and the image quality evaluation of real ultrasound carotid images. Further analysis of the performance of the above despeckle filters may be found in [1].

Table 1.2: An overview of despeckle filtering techniques presented in this book (see also vol. I [1])

SPECKLE REDUCTION TECHNIQUE	METHOD	INVESTIGATOR	FILTER NAME
Linear filtering	Moving window utilising local statistics		
	a) mean (m), variance (σ^2)	[6]-[19], [18]-[37]	*DsFlsmv*
	b) mean, variance, 3rd and 4th moments (higher statistical moments) and entropy	[6]-[19]	*DsFlsmvsk1d DsFlsmvsk2d*
	c) Homogeneous mask area filters	[38]	*DsFlsminsc*
			DsFwiener
	d) Wiener filtering	[19]-[21], [27]	
Non-Linear filtering	Median filtering	[39]	*DsFmedian*
	Linear scaling of the gray level values	[40]	*DsFls*
			DsFca DsFlecasort
	Based on the most homogeneous neighbourhood around each pixel	[26]	*DsFhomog*
	Non linear iterative algorithm (Geometric Filtering)	[15]	*DsFgf4d*
	The image is logarithmically transformed, the Fast Fourier transform (FFT) is computed, denoised, the inverse FFT is computed and finally exponentially transformed back	[21], [41], [42]	*DsFhomo*
	Hybrid median filtering	[43]	*DsFhmedian*
	Kuwahara filtering	[44]	*DsFKuwahara*
	Nonlocal Filtering	[45]	*DsFnlocal*
Diffusion filtering	Non-linear filtering technique for simultaneously performing contrast enhancement and noise reduction	[17], [18], [21], [24], [46]-[50],	*DsFad*
	Exponential damp kernel filters utilising diffusion	[24]	
	Speckle reducing anisotropic diffusion based on the coefficient of variation	[51]	*DsFsrad*
	Non-linear anisotropic diffusion	[51]	*DsFnldif*
	Non-linear complex diffusion	[52]	*DsFncdf*
Wavelet filtering	Threshold wavelet coefficients based on speckle noise at different levels	[37], [53]-[58]	*DsFwaveltc*

Table 1.3: Summary findings of despeckle filtering in an artificial carotid image (A), a phantom image (P), and a real ultrasound carotid image (C)

Despeckle Filter	Statistical and Texture Features	Measurements and shape features	Image Quality Evaluation	Optical Perception Evaluation
	A/P/C	A/P/C	A/P/C	A/P/C
Linear Filtering				
DsFlsmv	✓/✓/✓	-/✓/✓	✓/✓/✓	-/✓/✓
DsFlsminsc	-/-/✓	-/-/-	-/-/✓	-/-/-
Non-Linear Filtering				
DsFgf4d	-/-/-	-/-/-	-/-/-	✓/-/✓
DsFhomo	✓/-/-	-/-/-	-/-/-	-/-/-
DsFhmedian	✓/-/✓	✓/-/-	✓/✓/✓	✓/-/-
Diffusion Filtering				
DsFnldif	-/-/-	-/-/✓	-/-/✓	-/-/-
Wavelet Filtering				
DsFwaveltc		-/-/✓	-/-/✓	-/-/-

1.2   DESPECKLE FILTERING EVALUATION PROTOCOL

A despeckle filtering and evaluation protocol was documented in [1] and reproduced in Table 1.4 for easier reference. The despeckle filtering protocol given in Table 1.4 proposes the steps which are necessary to be taken when images will be processed for further image analysis. More specifically, following the acquisition of ultrasound images or videos, these are intensity normalised if needed and then despeckled. Texture features may then be extracted from the whole image or from an ROI and subsequently ROIs may be labelled or classified (i.e., normal vs abnormal tissue) depending on the application. The image quality metrics which are calculated between the original and the despeckled images may provide additional information of the level (i.e., no of iterations) of despeckle filtering as well as the quality of the despeckled image or video.

Two despeckle toolboxes supporting image and video processing respectively, that were developed by our team and are companion to this book are documented in Section 1.5. These toolboxes are freeware and can be used to investigate the usefulness of despeckle filtering in different problems under investigation.

1.3   SELECTED DESPECKLE FILTERING APPLICATIONS IN ULTRASOUND IMAGING AND VIDEO

There is a significant number of studies reported in the literature for despeckling of ultrasound images of the different imaging modalities presented in Table 1.1, using the different type of filters presented in Table 1.2. Table 1.5 tabulates selected despeckle filtering applications in ultrasound

Table 1.4: Despeckle filtering and evaluation protocol

Despeckle filtering and evaluation protocol
1	*Recording of ultrasound images*: Ultrasound images are acquired by ultrasound equipment and stored for further image processing. Regions of interest (ROI's) could be selected for further processing.
2	*Normalize the image*: The stored images may be retrieved and a normalized procedure may be applied (as described for example in section 2.2, companion monograph, Volume I [1]).
3	*Apply despeckle filtering*: Select the set of filters to apply despeckling together with their corresponding parameters (like moving window size, iterations, and other).
4	*Texture features analysis*: After despeckle filtering the user may select ROI's (i.e. the plaque or the area around the intima-media thickness) and extract texture features. Distance metrics between the original and the despeckled images may be computed (as well as between different classes of images if applicable).
5	*Compute image quality evaluation metrics*: On the selected ROI's compute image quality evaluation metrics between the original noisy and the despeckled images.
6	*Visual quality evaluation by experts*: The original and/or despeckled images may be visually evaluated by experts.
7	*Select the most appropriate despeckle filter/filters:* Based on steps 3 to 6 derive a performance evaluation table (see for example Table 1.3, Volume I [1]) and select the most appropriate filter(s) for the problem under investigation.

image covering the liver, pancreas, carotid artery, heart and others. A rather limited number of studies has been reported for video despeckling [45, 59]. The studies presented in Table 1.5 are grouped under the despeckle filtering groups already presented in Table 1.2.

In addition to the studies presented in Table 1.5 there is a plethora of studies published in the literature. A small number of these studies are discussed in this paragraph. Zain et al. [60] reported the use of average, median and Weiner filtering for speckle removal in ultrasound images of the liver. Senel et al. [61] applied the topological median filter to improve conventional median filtering on ultrasound images of the pancreas. The topological median filters implemented some existing ideas and some new ideas on fuzzy connectedness to improve the extraction of edges in noise (versus the conventional use of a median filter). A speckle reduction technique for 3D liver ultrasound images was proposed by extending the 2D speckle reducing anisotropic diffusion (SRAD) algorithm to a 3D algorithm [62]. The anisotropic diffusion model was effectively used for identifying edges in an image in the analysis phase, researched by Kim et al. [63]. Naderne-jad [64] used an anisotropic diffusion filter in artificial images where speckle noise was successfully

Table 1.5: An overview of selected despeckle filtering applications (*Continues*)

Principal Investigator	Method	Year	Input	N	Software Platform	Observers	Evaluation Metrics and Findings
Linear Filtering							
Frost [18]	Adaptive digital	1982	SAR, SIM	5	-	-	MSE=121
Lee [19]	Local Adaptive	1980	RWI	4	-	1	mean, σ^2=1.32
Lee [6], [19], [27]	Local statistics	1981	PUI, RWI, UI, SAR	4	C++	-	σ^2=0.82. Line plots
Kuan [16], [37]	Local statistics	1985	SAR	3	-	-	mean, σ^2
Nagao [38]	Homogeneous mask area	1979	SAR	3	-	-	mean=1.25, σ^2=1.27
Pizurica [66]	Generalized Likelihood	2003	MRI, PUI, UI	3	-	1	SNR=13.6
Burckhardt [21]	Wiener filtering	1978	UI	1	-	-	mean, σ^2
Nonlinear Filtering							
Zhan [67]	Nonlocal means	2014	LUI, PUI, CUI, SUI	10	M	1	MAE, COV, PSNR, SSIM Improved metrics and VI
Maggioni [68]	Nonlocal spatiotemporal transform (3D and 4D)	2012	RWV	8	M	1	PSNR=41, MOVIE index=0.94, VQA, σ, VI,
Buade [45]	Nonlocal means	2008	RWI, RWV, Field II	6	C++	-	SSIN=0.91
Lui [69]	Non-local means K-nearest neighbor	2010	RWV	5	M	1	σ=40, VI Intensity graphs
Chan [70]	Spatiotemporal varying	2005	RWV	5	-	-	PSNR=41.93
Varghese [71]	Gaussian scale mixture	2008	RWV	5	M	-	PSNR=41, SSIM=0.95 motion compensation
Biardar [72]	Fuzzy geometric Wiener	2014	CUI	5	M	2	PSNR=23, MSE=222, SSIM=0.99, ρ=0.91
Kuwahara [44]	Kuwahara filtering	1976	LUI	4	M	-	PSNR=27, RMSE, MSE=0.34
Coupe [73]	Nonlocal means	2009	LUI, MRI, PUI	4	-	-	SNR=64, QI=2.71
Zlokolica [74], [75]	α-Trimmed mean filter	2002	UV	3	C++	1	PSNR, VI, 3D filtering.
Busse [15]	Geometric Filter	1995	UI	3	-	-	mean=122,
Lupas [76]	Adaptive weighted median	1989	LUI	3	-	-	MSE=22
Niemen [41]	Hybrid median	1987	UI	3	-	1	PSNR=24, RMSE=0.043, MSE=20
Solbo [41], [42]	Homomorphic	2004	UI	3	-	-	ENL=22, mean=121, median=126, σ=1.4
Huang [39]	Median filtering	1979	RWI	2	F	-	Window size=5x5, 7x7, 9x9

Table 1.5: *(Continued)* An overview of selected despeckle filtering applications *(Continues)*

Principal Investigator	Method	Year	Input	N	Software Platform	Observers	Evaluation Metrics and Findings
Diffusion Filtering							
Cardoso [77]	Inference and anisotropic diffusion	2012	Filed II	4	M		ISFAD=90%, SSIM=0.95
Yongjian [24]	SRAD	2002	Field II, UCA	3	-		mean=122, σ^2=1.19, CV%=1.14, FOM=0.72
Narayanan [78]	Coupled PDE diffusion	2011	UI	2	-		CNR=2.56, SSIM=0.81, FOM=0.91
Perona-Malik [50]	Anisotropic Diffusion	1990	RWI	5	M	1	VI
Jin [46]	Nonlinear AD	2000	BUI, KUI,	4	-	-	Histograms
Elmoniem [51]	Nonlinear Coherent diffusion	2002	CUI, KUI, PUI, LUI	4	C++	-	Line plots, VI
Ullom [80]	Frequency compounding	2012	Field II	3	M	0	CNR=348%
Weickert [47]	Nonlinear diffusion	1998	MRI, PUI	2	-	2	CPU time=152 sec, %error=0.27%
Bernardes [52]	Complex diffusion	2010	PUI, UI	2	M	3	MSE=37, ENL=46.3, CNR=13.8, line plots
Tay [80]	Local Adaptive anisotropic diffusion	2010	Field II, CU	1	C++	1	SSIM=0.84
Llorden [81]	Memory Anisotropic diffusion	2015	KUI, IVUS	5	M	1	MS=0.001/SSIM=0.88/Q=3.21
Wavelet Filtering							
Zong [55]	Multiscale Wavelet	1998	CUI	10	-	-	MSE=29, Line plots
Abrahim [82]	Wavelet thresholding	2012	LUI	2	-	-	MSE=10, SNR=25, PSNR=45, SSIN=0.99, Q=0.85
Bioucas [83]	Variable splitting	2010	RWI	5	M	0	SNR=24, MAE=20, RE=17
Dabov [84]	Hard thresholding and Wiener (3D)	2007	RWV	5	M	1	PSNR=26
Rusanovsky [85]	Block matching and hard-thresholding	2006	RWV	3	C++	-	PSNR=34, VI
Gupta [86]	Multiscale (Wavelet)	2004	KUI	3	-	-	MSE=27, ρ=0.99, β=0.92 (shape parameter)
Yue [87]	Nonlinear multiscale	2006	LUI, UI	2	-	-	FOM=0.91, ρ=0.98, Line plots

Table 1.5: *(Continued)* An overview of selected despeckle filtering applications

Principal Investigator	Method	Year	N	Input	Software Platform	Observers	Evaluation Metrics and Findings
Review Studies							
Loizou [7]	Review	2005	440	Field II, UI	M	2	61 texture features,16 quality metrics, 10 despeckle, *DsFlsmv* best performance
Sivakumar [88]	Review	2010	200	KUI, LUI	M	1	21 quality metrics, 12 despeckle filters SRAD best performance
Zhang [89]	Review	2015	200	Field II, BUI	M	2	4 quality metrics, 11 despeckle filters. NIQE best quality metric
Finn [59]	Review	2011	100	Field II, UV,	-	1	5 quality metrics, 15 despeckle filters OSRAD best performance
Loizou [14]	Comparison	2012	10	UV	M	1	4 despeckle filters *DsFlsmv* best performance
Ortiz [90]	Review	2012	2	Field II, UI	M	-	3 quality metrics, 6 despeckle filters,
Biradar [91]	Review	2015	200	CUI	M	1	15 quality metrics, 10 despeckle filters, improved segmentation after filtering

BUI: Breast ultrasound imaging, CUI: Cardiac ultrasound images, IVUS: Intra vascular ultrasound images, KUI: Kidney ultrasound image, LUI: Liver ultrasound image, PUI: Phantom ultrasound images, RWI: Real world images, RWV: Real world videos, SAR: Synthetic aperture radar images, SIM: Simulated images, USC: Ultrasound muscular arteries, SUI: Spine ultrasound images, UCA: Ultrasound carotid artery images, UV: Ultrasound Video

AD: Anisotropic diffusion, β: Shape parameter, CV%: Coefficient of variation, CNR: Contrast to-noise ratio, ENL: Effective number of looks, FOSSIM: Structural similarity index, F: FORTRAN, M: Pratt's figure of merit, M: Matlab[®], MSE: Mean square error, MAE: Mean absolute error, N: Number of subjects investigated, OSRAD: Oriented speckle reducing anisotropic diffusion, PSNR: Peak signal-to-noise ratio, QI: Quality index, ρ: Correlation coefficient, RE: Relative error, SAR: Synthetic aperture radar, SNR: Signal-to-noise ratio, VI: Visual inspection.

removed. A comparative study between wavelet coefficient shrinkage filter and several standard despeckle filters showed that the discrete wavelet based filtering gave the best results for speckle removal [65].

1.4   SELECTED DESPECKLE FILTERING SOFTWARE

Table 1.6 tabulates selected ultrasonic image processing and analysis software systems. These systems are grouped under commercial and freeware imaging systems. The commercial software systems have no despeckle filtering functionality. There are several systems that are available in the market that support automated IMT measurement. These systems are included in the widely known commercial ultrasound machines (Esaote S.p.A, Philips Electronic Ltd) as well as the ones that can be purchased as stand-alone software systems (Segment™, Atheroedge™, Royal Perth IMT software and M'Ath®). More specifically, Davis et al. [92] patented the RF-QIMT & RF-QAS integrated commercial imaging software systems (Esaote S.p.A, Florence, Italy) for the segmentation and measurement of IMT. This software was evaluated on 635 subjects. The Segment™ software used by Phillips Electronics Ltd was also presented by Heiberg et al. [94] and supports cardiovascular image analysis. Molinari and co-workers [97] published an automated system which segments the lumen-intima and the media-adventitia borders in 2D ultrasound images and then measures the IMT. This system was patented by Global Biomedical Technologies, Inc., CA, USA, under the name AtheroEdge™. A commercial IMT software system tool that can be used to measure the IMT and the lumen diameter in ultrasound images, and image sequences of the CCA was developed by the Royal Perth Hospital, Australia [92]. Moreover, a software system from M'Ath®Intelligent Medical Technologies, Argenteul, France was presented for the segmentation of the IMC and measurement of the IMT from 2D ultrasound images of the CCA. Furthermore, Kyriakou et al. [95] introduced a commercial software system supporting the automated image normalization, manual delineation, texture and risk analysis of the atherosclerotic carotid plaque in ultrasound images of the CCA.

     Freeware imaging systems were developed by Xu et al. [93] and Loizou and co-workers [3, 7, 8, 11]. The software by Xu et al. [93], supports different imaging formats, 3D reconstruction, processing and visualization. The software by Loizou and co-workers for image [3, 7, 8, 11] and video [2], supports normalization, despeckle filtering, texture analysis, image quality evaluation and manual delineation as documented in the following section where the IDF and VDF freeware despeckling software toolboxes are introduced.

1.5   THE IMAGE AND VIDEO DESPECKLE FILTERING TOOLBOXES

In this section the IDF and VDF image and video despeckling software systems developed by our group are briefly introduced. More details about their functionality and link for downloading are given in the Appendices A.1 and A.2, respectively.

Table 1.6: An overview of selected ultrasound common imaging software systems

Principal Investigator	Year	Method	2D/3D	Software Platform	N	System Features
Commercial Imaging Systems						
Xu [94]	2008	Ultra3D®, 3D Analysis	3D	Windows	-	TIF, JPG, PNG, image acquisition, 3D image reconstruction, image pre-processing, 3D visualization.
Segment™, Phillips Electronics Ltd [95]	2010	Cardiovascular Analysis	2D/3D	Windows/ Matlab®	-	DICOM, raw, TIF, JPG, ROI manual and automated segmentation.
Kyriakou [96]	2007	Normalization and Plaque Texture Analysis	2D	Matlab®	440	Supports all image types. Manual segmentation, image normalization, texture analysis.
Freeware Imaging Systems-IMC measurement						
Usimagtool® [97]	2007	Image Analysis	2D/3D	C++	-	DICOM, raw, TIF, JPG, VTK. Additive filtering, edge detection, automated segmentation, volume visualization.
Loizou, [7], [8], [11], [2], [3]	2005, 2006, 2008 2014	Despeckle Image Filtering	2D	Matlab®	440	TIF, JPG, PNG, Matlab® files. Manual, automated segmentation, image normalization, selection of despeckle filtering, texture and image quality analysis. ROI processing.

N: Number of cases investigated.

We have proposed in [1, 2], an integrated despeckle filtering (IDF) software toolbox (see also Fig. 1.1 and Fig. 1.2) for preprocessing of ultrasound images. The work presented in [1, 2], incorporates results also presented in previous publications made by our group, where despeckle filtering [7, 26], segmentation of the intima-media complex [98], the atherosclerotic carotid plaque [9], and image quality evaluation [8], from ultrasound images of the CCA were investigated.

Using the IDF toolbox 15 different despeckle filtering techniques were evaluated on 100 ultrasound images of the CCA. Furthermore, the performance of despeckle filtering was evaluated on 65 different texture features and 15 image quality evaluation metrics using the IDF toolbox. These were extracted from the original and the despeckled images well as evaluated by visual perception, performed by the experts.

The VDF integrated software toolbox [1, 3] implements procedures for loading video files of the most common formats (see also Fig. 1.3). The video analysis usually begins with prescribing an ROI in the first video frame where despeckle filtering is applied in a frame by frame basis for the whole video sequence. There are 65 different texture features [1] and six different video quality metrics that are extracted and evaluated by comparing the original and despeckled videos. Figure 1.1 presents a flowchart of the analysis of the proposed integrated VDF system, where the different modules of the software are outlined. In Fig. 1.3 we present the GUI of the proposed system. The VDF toolbox for medical ultrasound video was proposed in [1, 3, 14] and evaluated on 20 ultrasound videos of the CCA.

1.6   GUIDE TO BOOK CONTENTS

In the following chapter, the application and results of the segmentation of the intima-media complex (IMC), the media-layer (ML) and the intima layer (IL) of the common carotid artery as well as the segmentation of the atherosclerotic carotid plaque from ultrasound images and videos following despeckle filtering are presented. In Chapter 3, we present the results on image and video texture analysis. We provide results from the texture analysis of the IMC and the atherosclerotic carotid plaque performed on a large number of ultrasound images and videos. In Chapter 4 we present results on ultrasound wireless video transmission and encoding which are performed before and after despeckle filtering. Chapter 5 discusses, compares and evaluates the proposed despeckle filtering techniques for image and video and provides an outline of future directions. Finally, at the end of this book, appendices provide details about the IDF and VDF despeckle filtering MATLAB™ toolboxes.

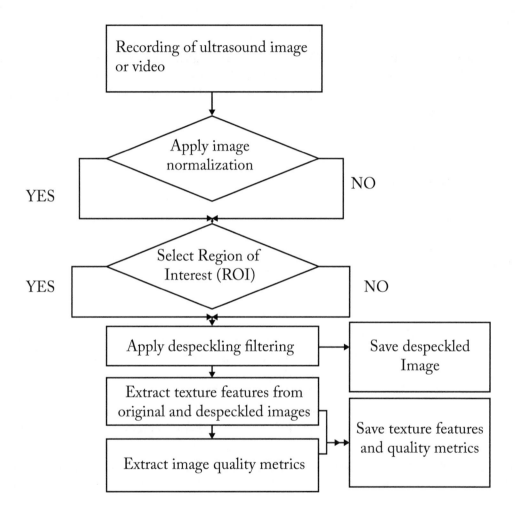

Figure 1.1: Flowchart analysis of the IDF/VDF toolboxes for ultrasound image/video analysis.

Figure 1.2: The graphical user interface (GUI) of the IDF toolbox [2] for ultrasound image analysis when filtering is applied in an ROI selected by the user of the system. The following components are shown: original image settings, original image display, filter options, despeckled image display and evaluation metrics for texture and quality.

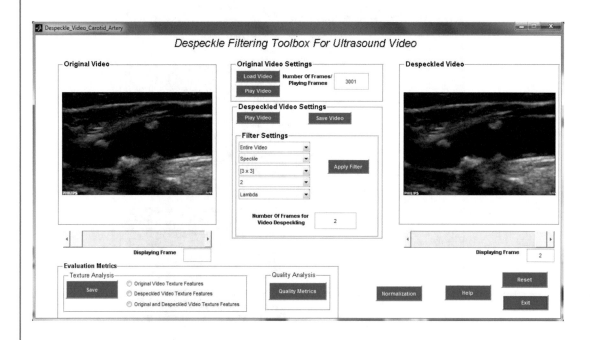

Figure 1.3: The GUI of the despeckle filtering toolbox for ultrasound video (VDF) [3]. The following components are shown: original video display, original video settings, despeckled video display, despeckled video settings (filter settings), number of frames for video despeckling, and evaluation metrics for texture analysis and quality analysis.

CHAPTER 2

Segmentation of the Intima-media Complex and Plaque in CCA Ultrasound Imaging and Video Following Despeckle Filtering

In the following chapters we will present the application and results of the segmentation of the IMC the ML and the IL of the CCA as well as the segmentation of the atherosclerotic carotid plaque from ultrasound images and videos.

Ultrasound measurements of the human carotid artery walls are conventionally obtained by manually tracing interfaces between tissue layers. In this section we present results from a snakes segmentation technique [98, 99] for detecting the intima-media layer the ML and IL [99], of the far wall of the CCA in longitudinal ultrasound images (see Fig. 2.1), by applying snakes [99]–[101], after normalization, speckle reduction, and normalization and speckle reduction. The IMT (see Fig. 2.1) of the CCA can serve as an early indicator of the development of cardiovascular disease, like myocardial infarction and stroke [102]. Previous studies indicated that increase in the IMT of the CCA is directly associated with an increased risk of myocardial infarction and stroke, especially in elderly adults without any history of cardiovascular disease [102, 103]. Importantly, increased IMT was demonstrated to have a strong correlation with the presence of atherosclerosis elsewhere in the body and may thus be used as a descriptive index of individual atherosclerosis [104]. As vascular disease develops, local changes occur in arterial structure, which thicken the innermost vessel layers known as IMC. As disease progresses the IMT initially increases diffusely along the artery and then becomes more focal, forming discrete lesions or plaques, which gradually grow and obstruct blood flow. Furthermore, these plaques can become unstable and rupture with debris transported distally by blood to obstruct more distal vessels. This is particular so if plaques develop internal pools of lipid covered only by a thin fibrous cap [104]. It is therefore important to accurately estimate the IMT.

It was proposed but not thoroughly investigated, that not only the IMT but rather the ML (its thickness and its textural characteristics) may be used for evaluating the risk that a sub-

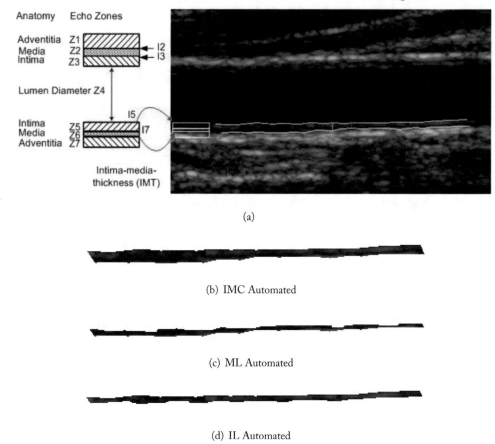

(a)

(b) IMC Automated

(c) ML Automated

(d) IL Automated

Figure 2.1: a) Illustration of the intima-media-complex (IMC, bands Z5 and Z6) of the far wall of the common carotid artery and the automatic IMC segmentation. The IMC consists of the intima band (Z5), the media band (Z6) and the far wall adventitia band (Z7). The IMT complex is defined as the distance between the blood intima interface line and the media adventitia interface line. The media layer (ML) is defined as the layer (band) between the intima-media and the media-adventitia interface (band Z6), (c) extracted IMC, (d) extracted media layer (ML) and (e) extracted intima layer (IL). Source [101] © CMIG 2009.

ject might develop stroke. As shown in Fig. 2.1, the IL is a thin layer, the thickness of which increases with age, from single cell layer at birth to 250 μm at the age of 40 for non-diseased individuals [105]. In ultrasound images, the media layer (ML) is characterized by an echolucent region, predominantly composed of smooth muscle cells, enclosed by the intima and adventitia layers (see Fig. 2.1, band Z6) [106]. Earlier research [107], showed that the media layer thickness (MLT) does not change significantly with age (125 μm <MLT< 350 μm). In [99], the median

(IQR) of intima layer thickness (ILT), MLT, and IMT were computed from 100 ultrasound images of 42 female and 58 male asymptomatic subjects aged between 26 and 95 years old, with a mean age of 54 years to be as follows 0.43 mm (0.12), 0.23 mm (0.18), and 0.66 mm (0.18), respectively.

The use of ultrasound provides a non-invasive method for estimating the IMT of human carotid arteries and is especially suited to dynamic analysis owing to its ability to deliver real-time video sequences. A B-mode ultrasound image shown in Fig. 2.1, shows the IMC at the far wall of the carotid artery, (echo zones Z5-Z6), as a pair of parallel bands, an echodense and an echolucent. The band Z5 and the leading edge of the band Z7 (adventitia) denoted as I5 and I7 define the far-wall IMT. With this understanding, the determination of the IMT at the far wall of the artery becomes equivalent to accurately detecting the leading echo boundaries I5 and I7. The lumen-intima and media-adventitia intensity interface of the far wall of the carotid artery is preferred for IMT measurements [98, 108]. It has been shown that the definition of the IMT as illustrated in Fig. 2.1 corresponds to the actual histological IMT [98, 102].

Traditionally, the IMT is measured by manual delineation of the intima and the adventitia layer [98, 103, 108]. Manual tracing of the lumen diameter (see Fig. 2.1 Z4) and the IMT (see Fig. 2.1 I5, I7) by human experts requires substantial experience, it is time consuming and varies according to the training, experience and the subjective judgment of the experts. The manual measurements suffer therefore from considerable inter- and intra-observer variability [98, 103, 108].

Carotid artery atherosclerosis is the primary cause of stroke and the third leading cause of death in the United States. Almost twice as many people die from cardiovascular disease than from all forms of cancer combined. Atherosclerosis is a disease of the large and medium size arteries, and it is characterized by plaque formation due to progressive intimal accumulation of lipid, protein, and cholesterol esters in the blood vessel wall [109], which reduces blood flow significantly. The risk of stroke increases with the severity of carotid stenosis and is reduced after carotid endarterectomy [110]. The degree of internal carotid stenosis is the only well-established measurement that is used to assess the risk of stroke [111]. Indeed, it is the only criterion at present used to decide whether carotid endarterectomy is indicated or not [112].

2.1 SEGMENTATION OF THE IMC, ML AND IL IN ULTRASOUND IMAGING AND VIDEO

2.1.1 METHODOLOGY FOR THE SEGMENTATION OF THE IMC, ML AND IL IN ULTRASOUND IMAGING

A total of 100 B-mode longitudinal ultrasound images of the CCA used for the IMC, ML, and IL segmentations were recorded and the details are given in Appendix A.4 (see Database 3).

Brightness adjustments of ultrasound images were carried out in this study based on the method introduced in [5, 11]. It was shown that this method improves image compatibility by reducing the variability introduced by different gain settings, different operators, different equipment, and facilitates ultrasound tissue comparability. Algebraic (linear) scaling of the image was performed by linearly adjusting the image so that the median gray level value of the blood was 0–5, and the median gray level of the adventitia (artery wall) was 180–190 [5]. The scale of the gray level of the images ranged from 0–255. Thus, the brightness of all pixels in the image was readjusted according to the linear scale defined by selecting the two reference regions. It is noted that a key point to maintaining a high reproducibility was to ensure that the ultrasound beam was at right angles to the adventitia, adventitia was visible adjacent to the plaque and that for image normalization a standard sample consisting of the half of the width of the brightest area of adventitia was obtained.

Following image normalization, speckle reduction filtering, using the IDF toolbox [2] was applied to the region of interest uning the *DsFlsmv* despeckle filter with a moving sliding window of 5×5 and 4 itterations [1].

A neurovascular expert delineated manually (using the mouse) the IMC, the ML and the IL on the 100 longitudinal ultrasound images of the CCA after image normalization and speckle reduction filtering [1]. The IMC was measured by selecting 20–40 consecutive points for the intima (see Fig. 2.1a interface I5) and the adventitia (see Fig. 2.1 interface I7) layers, and the ML (see Fig. 2.1, band Z6) by selecting 10–20 consecutive points for the media (see Fig. 2.1, interface I6) and the adventitia layers at the far wall of the CCA. The IL (see Fig. 2.1, band Z5) was then derived by subtracting the ML from the IMC. The manual delineations were performed using the IDF toolbox [2]. The measurements were performed between 1–2 cm proximal to the bifurcation of the CCA on the far wall [99]–[101] over a distance of 1.5 cm starting at a point 0.5 cm and ending at a point 2.0 cm proximal to the carotid bifurcation. The bifurcation of the CCA was used as a guide and all measurements were made from that region. The IMT, MLT and the ILT were then calculated as the average of all the measurements. The measuring points and delineations were saved for comparison with the snakes segmentation method. All sets of manual segmentation measurements were performed by the expert in a blinded manner, both with respect to identifying the subject and as to the image delineation.

The 100 ultrasound images of the CCA were segmented to identify IMC, ML, and IL. Segmentation was carried out after image normalization using the automated snakes segmentation system proposed and evaluated on ultrasound images of the CCA in [98, 99], which is based on the Williams & Shah [114] snake. Initially the IMC was segmented by a snake segmentation system as proposed in [98, 99], where the boundaries I5 (lumen-intima interface) and I7 (media-adventitia interface) were extracted. Details about the implementation of the algorithm can be found in [11, 98, 99].

The upper side of the ML (see Fig. 2.1, Z6) was estimated by deforming the lumen-intima interface (boundary I5) by 0.36 mm (6 pixels) downwards and then deformed by the snakes segmentation algorithm proposed in [98] in order to fit to the media boundary (see Fig. 2.1a, interface I6). This displacement of 0.36 mm is based on the observation that the manual mean IMT (IMT$_{mean}$) is 0.71 mm (12 pixels), and lies between 0.54 mm (minimum or 9 pixels) and 0.88 mm (maximum or 15 pixels) [98, 99]. Therefore, the displacement of the contour, in order to estimate the media should be in average 0.36 mm (6 pixels times 0.06 mm) downwards, which is the half of the size of the IMT (the distance between I5 and I7, where I7 is the media-adventitia interface).

In order to achieve standardization in extracting the thickness of IMC, ML, and IL, segments with similar dimensions were divided based on the following procedure. A region of interest of 9.6 mm (160 pixels) in length, located 1–2 cm proximal to the bifurcation of the CCA, on the far wall was extracted. This was done by estimating the center of the IMC area and then selecting 4.8 mm (80 pixels) left and 4.8 mm (80 pixels) right of the center of the segmented IMC. The selection of the same ML area in length from each image is important in order to be able to make comparable measurements between images and patient groups. The novelty of the proposed methodology lies in the algorithmic integrated approach that facilitates the automated segmentation and measurements of IL and ML.

Following the segmentation of the mean, and median values for the IMT (IMT_{mean}, IMT_{median}, IMT_{max}, IMT_{min}), the MLT (MLT_{mean}, MLT_{median}), and the ILT (ILT_{mean}, ILT_{median}, ILT_{max}, ILT_{min}), and the inter-observer error for the IMT, MLT, and the ILT ($se = \sigma/\sqrt{2}$) [11, 134]. We also calculated the coefficient of variation, CV%, for the IMT, MLT and ILT respectively, which describes the difference as a percentage of the pooled mean values, where for the media $CV\%_{Median} = \frac{se_{Median}*100}{MLT_{median}}$ [11, 134].

The Wilcoxon rank sum test, which calculates the difference between the sum of the ranks of two dependent samples, was also used in order to identify if a significant difference (S) or not (NS) exists at $p < 0.05$, between the manual and the snakes segmentation measurements, of IMC, ML, IL for all 100 images, and between the manual and automated segmentation measurements.

2.1.2 METHODOLOGY FOR THE SEGMENTATION OF THE IMC IN ULTRASOUND VIDEO

A total of 10 B-mode longitudinal ultrasound videos of the CCA which display the vascular wall as a regular pattern that correlates with anatomical layers were recorded for the IMC segmentation. The videos were acquired by the ATL HDI-5000 ultrasound scanner (Advanced Technology Laboratories, Seattle, USA) [186] with a resolution of 576×768 pixels with 256 gray levels, a spatial gray resolution of 17 pixels per mm (i.e., the resolution is 60 μm) and having a frame rate of 100 frames/sec. All video frames were manually resolution-normalized at 16.66 pixels/mm. This was carried out to overcome the small variations in the number of pixels per mm of image depth (i.e., for deeply situated carotid arteries, image depth was increased and therefore digital image spatial resolution would have decreased) and in order to maintain uniformity in the digital image spatial resolution [5]. The videos were recorded at the Saint Mary's Hospital, Imperial College of Medicine, Science and Technology, UK, from asymptomatic patients.

Brightness adjustments of ultrasound videos were carried out in this study based on the method introduced in [5], which improves image compatibility by reducing the variability introduced by different gain settings, different operators, different equipment, and facilitates ultrasound tissue comparability. Additonal details on the normalization procedure were also given in earlier chapter and can aslo be found in [1, 3, 13, 14] and [118].

The despeckle *DsFlsmv* filter for video despeckling, which was introduced in [1, 14] was applied to each consecutive frame prior to IMC segmentation. The filter was applied for 2 itterations to each video frame using a 5×5 pixel moving window.

An expert (vascular surgeon) manually delineated the IMC borders every 20 frames on 10 longitudinal B-mode ultrasound videos of the CCA. The three contours correspond to the far wall media-adventitia interface, the far wall lumen-intima interface and the near wall intima-lumen interface (see Fig. 2.3a). This procedure was carried out after image normalization and speckle reduction filtering using the VDF toolbox [14] developed by our group.

The automated segmentation procedure is described in detail in [98]. Measurements were carried out for 3–5 seconds, covering in general 2–3 cardiac cycles, 1 cm proximal to the bifurcation. The manual IMT was measured in the first frame and then every 20 frames per subject over the whole cardiac cycle. The automated IMT was measured at each frame per subject over the whole cardiac cycle. The distance was computed between the two boundaries, at all points along the arterial segment of interest moving perpendicularly between pixel pairs, and then averaged to obtain the mean IMT (IMT_{mean}). Also the maximum (IMT_{max}), minimum (IMT_{min}), and median (IMT_{median}) IMT values, are calculated. Figure 2.3a shows the detected IMT_{mean} on the first video frame.

In order to investigate how the results of the snakes segmentation method differ from the manual delineation results, we used the following measurements. We computed the parameters IMT_{mean}, IMT_{min}, IMT_{max}, and IMT_{median}, for each video frame as well as the standard deviation over the whole video frames at end diastole. We have also computed the carotid diameter during contraction (CDC), the carotid diameter during distension (CDD) and the percentage of the carotid wall distension (%CWD=((CDD-CDC)/CDC)*100%). Furthermore, in order to evaluate our algorithm, the following evaluation metrics between the automated and manual contours for all three detected borders were computed. The three contours correspond to the far wall media-adventitia interface, the far wall lumen-intima interface and the near wall intima-lumen interface (see Fig. 2.3).

2.1.3 RESULTS OF THE SEGMENTATION OF THE IMC, ML AND IL IN ULTRASOUND IMAGING

Figure 2.2 shows a longitudinal ultrasound image of the CCA with the manual delineations from the two experts (Fig. 2.2b, Fig. 2.2c), the automatic initial contour estimation (Fig. 2.2d), and the Williams & Shah snakes segmentation results for the cases of no pre-processing (NP) (Fig. 2.2e), despeckled (DS) (Fig. 2.2f), normalized (N) (Fig. 2.2g), and normalized despeckled (NDS) (Fig. 2.2h). The detected IMT_{mean}, IMT_{max}, and IMT_{min} values, are shown with a double, single, and dashed line boxes respectively. The results in Fig. 2.2 showed, that the IMT was

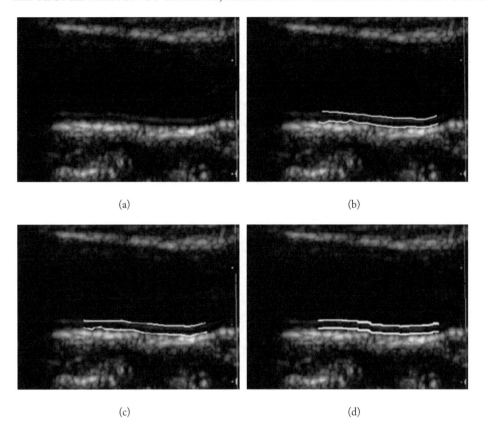

Figure 2.2: (a) Original longitudinal ultrasound image of the carotid artery, (b) manual delineation from Expert 1, (c) manual delineation from Expert 2, (d) initial contour estimation. Source [98], © MBEC 2007. *(Continues.)*

detected in all snakes segmentation measurements but with variations between experts and methods.

The manual (for two different experts, Expert 1, Expert 2) and the automated IMT_{mean}, IMT_{min}, IMT_{max}, and IMT_{median}, as well as the MLT_{mean}, MLT_{min}, MLT_{max}, and MLT_{median}, and the ILT_{mean}, ILT_{min}, ILT_{max}, and ILT_{median}, measurements for Fig. 2.2 are presented in Table 2.1. The manual measurements are given for each expert (Expert 1 and Expert 2), in cases when manual measurements were carried out, without normalization (M) and with normalization (MN). The Williams & Shah snakes segmentation [114] measurements are given for the NF, DS, N and NDS cases, and were in the most of the cases, higher than the manual measurements, except in the MN case for both experts. The observed standard deviation, *sd*, values for the IMT_{mean}, was for the first expert, M (0.14), MN (0.11), for the second expert, M (0.12), MN (0.15), and for the

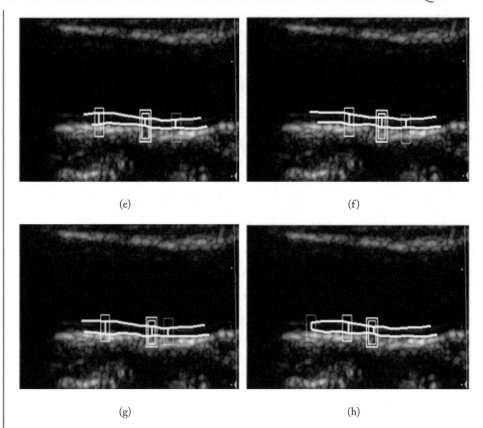

(e)   (f)

(g)   (h)

Figure 2.2: *(Continued.)* The segmentation results of the IMT for (e) no filtering (NF), (f) despeck-led (DS), (g) normalized (N), and (h) normalized despeckled (NDS) images. The detected IMT_{mean}, IMT_{max}, and IMT_{min} are shown with a double, single, and dashed line boxes, respectively. Source [98], © MBEC 2007.

snakes segmentation, NF (0.22), DS (0.21), N (0.19), and NDS (0.18), respectively. The results in Fig. 2.2 and Table 2.1 show, that the IMT, MLT and ILT were detected well in all snakes segmentation measurements but with variations between experts and methods. The best visual results as assessed by the two vascular experts were obtained on the NDS, followed by N and DS images.

Table 2.2 tabulates the first set of manual and the Williams & Shah snakes segmentation results for 100 longitudinal ultrasound images of the carotid artery, for the IMT_{mean}, IMT_{min}, IMT_{max}, and IMT_{median}, with their standard deviations (sd), the interobserver error (se), and the coefficient of variation (CV%). The $IMT_{mean} \pm sd$ results for Expert 1 were, 0.67 ± 0.16 mm, 0.68 ± 0.17 mm, and for Expert 2 were, 0.65 ± 0.18 mm, 0.61 ± 0.17 mm on the NP and N

Table 2.1: Comparison between the manual and the snakes IMT, MLT and ILT segmentation measurements for the cases (b)–(h) in Fig. 2.2. Measurements are in millimeters (mm)

	Manual Measurements				Snakes Segmentation Measurements			
	Expert 1		Expert 2					
	M	MN	M	MN	NF	DS	N	NDS
IMT Measurements								
IMT_{mean}	0.74	0.92	0.82	0.98	0.82	0.81	0.82	0.82
(sd)	(0.14)	(0.11)	(0.12)	(0.15)	(0.16)	(0.18)	(0.17)	(0.18)
IMT_{min}	0.38	0.76	0.71	0.72	0.61	0.60	0.60	0.60
IMT_{max}	0.95	1.05	0.94	1.10	1.09	1.08	1.08	1.08
IMT_{median}	0.66	0.90	0.85	0.95	0.79	0.78	0.78	0.78
ML Measurements								
MLT_{mean}	0.29	0.31	0.32	0.34	0.27	0.26	0.26	0.26
(sd)	(0.13)	(0.11)	(0.14)	(0.14)	(0.15)	(0.15)	(0.16)	(0.15)
MLT_{min}	0.11	0.11	0.13	0.12	0.11	0.11	0.11	0.11
MLT_{max}	0.31	0.33	0.33	0.36	0.32	0.33	0.32	0.33
MLT_{median}	0.27	0.26	0.29	0.22	0.25	0.25	0.26	0.27
IL Measurements								
ILT_{mean}	0.45	0.61	0.5	0.64	0.55	0.55	0.56	0.56
(sd)	(0.11)	(0.12)	(0.13)	(0.15)	(0.16)	(0.17)	(0.16)	(0.17)
ILT_{min}	0.27	0.65	0.58	0.60	0.50	0.49	0.49	0.49
ILT_{max}	0.64	0.72	0.61	0.74	0.77	0.75	0.76	0.75
ILT_{median}	0.39	0.64	0.56	0.73	0.54	0.53	0.52	0.51

M: Manual, MN: Manual normalised, NF: No filtering, DS: Despeckle, N: Normalised, NDS: Normalized despeckled, sd : Standard deviation. Sources [101] © CMPB 2009, [98] © MBEC 2007, [99] © IEEE TUFFC.

images, respectively, [98]. The $IMT_{mean} \pm sd$ snakes segmentation results were 0.7 ± 0.14 mm, 0.69 ± 0.13 mm, 0.67 ± 0.13 mm, 0.68 ± 0.12 mm, for the NF, DS, N, and NDS images, respectively. The middle part of Table 2.2 presents the second set of manual measurements for the 100 images of the carotid artery made by the two experts 12 months after the first set of measurements were made. This was carried out in order to assess the intra-observer variability for the same expert. It is shown that the measurements made by Expert 2 are generally smaller giving a thinner IMT. Furthermore, the sd, the se, and the CV%, for the measurements made by Expert 2 are also smaller.

The results from Table 2.2 showed that high intra observer variabilities occur when manual measurements are made. It is documented in the literature that measurements of the se, can be used as clinically useful standard to measure the performance of image segmentation algorithms [115]. There are some results given in the literature for the intra observer variability for the IMC segmentation performed in carotid artery images. Specifically, in [116] the $IMT_{mean} \pm sd$ results of Expert 1 and Expert 2 were 0.87 ± 0.12 mm and 0.90 ± 0.2 mm, respectively. In this study for the second set of measurements the results of Expert 1 and Expert 2 were 0.85 ± 0.11 mm

Table 2.2: Comparison between manual and snakes segmentation measurements for 100 ultrasound images of the CCA. Measurements are in millimeters (mm). Bolded values show best performance.

| | Manual first set of measurements at time 0 | | | | Manual second set of measurements at time 12 months | | | | Snakes segmentation measurements | | | |
| | Expert 1 | | Expert 2 | | Expert 1 | | Expert 2 | | | | | |
	NF	N	NF	N	NF	N	NF	N	NF	DS	N	NDS
IMT_{mean}	0.67	0.68	0.65	0.61	0.67	0.68	0.55	0.57	0.70	0.69	0.67	**0.68**
(sd)	(0.16)	(0.17)	(0.18)	(0.17)	(0.16)	(0.17)	(0.11)	(0.13)	(0.14)	(0.13)	(0.13)	**(0.12)**
IMT_{min}	0.53	0.52	0.57	0.54	0.53	0.52	0.45	0.47	0.51	0.51	0.51	**0.49**
(sd)	(0.14)	(0.15)	(0.16)	(0.14)	(0.14)	(0.15)	(0.11)	(0.14)	(0.13)	(0.13)	(0.14)	**(0.11)**
IMT_{max}	0.82	0.85	0.75	0.70	0.82	0.85	0.64	0.66	0.90	0.88	0.86	**0.87**
(sd)	(0.22)	(0.21)	(0.19)	(0.20)	(0.22)	(0.21)	(0.13)	(0.14)	(0.20)	(0.19)	(0.17)	**(0.15)**
IMT_{median}	0.66	0.66	0.67	0.61	0.66	0.66	0.62	0.61	0.69	0.69	0.66	**0.64**
(sd)	(0.16)	(0.18)	(0.18)	(0.17)	(0.16)	(0.18)	(0.16)	(0.14)	(0.14)	(0.13)	(0.12)	**(0.12)**
se	0.11	0.12	0.13	0.11	0.11	0.12	0.08	0.1	0.10	0.09	0.09	**0.08**
$CV\%$	16.7	17.1	19.1	17.2	16.7	17.1	14.0	16.8	13.8	13.4	13.2	**12.5**

NF: No pre-processing, N: Normalized, DS: Despeckled, NDS: Normalized despeckled, sd : Standard deviation, se : Intra-observer error for mean values, $CV\%$: Coefficient of variation. Source [98] © MBEC 2007.

and 0.85 ± 0.17 mm, respectively. It should be noted that direct comparisons between different studies, are difficult, due to the dependence on the measurement protocol, number and type of patients, tissue to be segmented and image quality.

Table 2.3 shows the results of the Wilcoxon rank sum test, and a variation of the Haussdorf distance (HD) between Expert 1 and Expert 2 and the snakes segmentation measurements. The Wilcoxon rank sum test, which is displayed in the upper triangle of the left and right columns of Table 2.3, showed that NS difference existed between the Williams & Shah snakes segmentation measurements and the manual measurements from Expert 1 and Expert 2, respectively. The NS difference between the two methods showed that the manual measurements may be replaced by the snakes segmentation measurements with confidence. The HD displayed in the left and right columns lower triangle of Table 2.3, showed that minimum mismatches were obtained for Expert 1, between the N-Manual first set NF (3.4), NDS-Manual first set N (4.7), NDS-N (5), DS-NF (5.2), NDS-Manual first set NF and DS-Manual first set N (8.6), and for Expert 2 between the NDS-N (5), and DS-NF (5.2), respectively.

Table 2.3 showed that NS differences between the manual and the Williams & Shah snakes segmentation method for all manual and automated segmentation cases (NF, DS, N, NDS) were found using the Wilcoxon rank sum test. The smallest HD, (see Table 2.3), was found between the N-Manual first set NF (3.4), NDS- Manual first set NF (4.7), NDS-N (5), and DS-NF (5.2) images, which showed the minimum mismatches between these measurements.

2.1.4 RESULTS OF THE SEGMENTATION OF THE IMC IN ULTRASOUND VIDEO

The analysis of the arterial wall thickening over the entire cardiac cycle may provide additional clinical information for the atherosclerosis disease. In addition to the IMT estimation, changes in the mechanical properties of the arterial wall are of interest because they also have the potential to indicate the existence of early cardiovascular diseases. These changes can be detected by analyzing the arterial wall stiffness (or elasticity) using techniques such as diameter change estimation, artery distensibility, or strain imaging [117].

Figure 2.3 illustrates an example of a CCA video IMT and carotid diameter manual (see Fig. 2.3, left column) and automated (see Fig. 2.3, right column) segmentations at, the 1st, 50th, 100th and 150th frames of video. It is shown that, manual and automated segmentation measurements are visually very similar.

Figure 2.4 illustrates the lumen rate of change of a video of a CCA, showing the diameter change over all video frames, while Fig. 2.4b presents the IMT rate of change over all video frames for one cardiac cycle. It is shown that the IMT has a variation of 0.02 mm over all cardiac cycles, with a mean IMT of 0.84 mm, while the diameter ranges between 6.9 mm and 7.63 mm, with a carotid diameter difference of 0.73 mm.

Table 2.4 illustrates the results for the mean, median, minimum and maximum IMT values estimated manually and automatically from the proposed algorithm for all videos investi-

Table 2.3: Tests and measures computed on 100 ultrasound images of the carotid artery from Expert 1 and Expert 2 and the snakes segmentation measurements. Left and right columns upper triangle: Wilcoxon rank sum test with the p value shown in parentheses. Left and right columns lower triangle: Variation of the HD($*10\text{-}3$ mm). Bolded values show best performance.

	Wilcoxon rank sum test and HD												
	First set manual Expert 1		Automated					First set manual Expert 2			Automated		
	NF	N	NF	DS	N	NDS		NF	N	NF	DS	N	NDS
NF	-	-	NS (0.45)	NS (0.56)	NS (0.64)	NS (0.9)	NP	-	-	NS (0.06)	NS (0.1)	NS (0.07)	NS (0.09)
N	13.3	-	NS (0.90)	NS (0.79)	NS (0.30)	NS (0.55)	N	40	-	NS (0.08)	NS (0.07)	NS (0.1)	NS (0.16)
NF	27.1	13.8	-	NS (0.87)	NS (0.33)	NS (0.53)	NP	46.2	86.2	-	NS (0.87)	NS (0.33)	NS (0.53)
DS	21.9	**8.6**	**5.2**	-	NS (0.45)	NS (0.69)	DS	41	81	**5.2**	-	NS (0.41)	NS (0.55)
N	**3.4**	9.9	23.7	19	-	NS (0.41)	N	22.5	62.5	23.7	19	-	NS (0.69)
NDS	**8.6**	**4.7**	18.5	13	5	-	NDS	27.7	67.7	18.5	13	5	-

NF: No pre-processing, N: Normalized, DS: Despeckled, NDS: Normalized despeckled. The p value is also shown in parentheses (S=Significant difference at $p \leq 0.05$, NS=Non significant difference at $p > 0.05$). Source [99] © MBEC 2007.

Manual Segmentation   Snakes base Segmentation

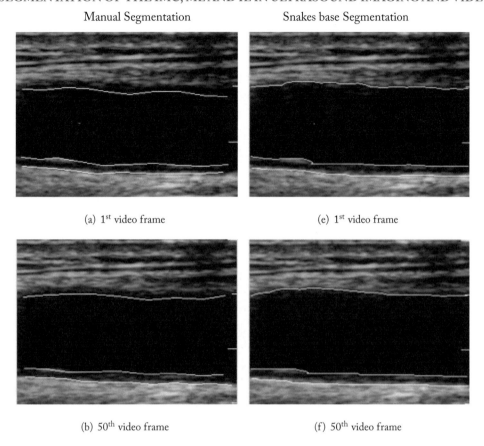

(a) 1st video frame   (e) 1st video frame

(b) 50th video frame   (f) 50th video frame

Figure 2.3: Manual (left column) and automated (right column) video segmentation of the CCA for the IMT and carotid diameter, at the (a) 1st ($IMT_{man} = 0.85$ mm, $diameter_{man} = 6.69$ mm, $IMT = 0.99$ mm, $diameter = 6.9$ mm), (b) 50th ($IMT_{man} = 0.90$ mm, $diameter_{man} = 7.31$ mm, $IMT = 0.93$ mm, $diameter = 7.51$ mm). Source [118], © IEEE 2012. *(Continues.)*

gated. The results after the automated segmentation of 10 videos of the CCA gave values of $IMT_{mean} = 0.72 \pm 0.4$ mm, $IMT_{median} = 0.71 \pm 0.21$ mm, $IMT_{min} = 0.67 \pm 1.1$ mm, $IMT_{mean} = 0.76 \pm 0.21$ mm. No significance difference was found between the manual and the automated IMT segmentation measurements ($p = 0.003$).

Table 2.5 presents the manual and automated results of the carotid diameter during contraction (CDC), the carotid diameter during distension (CDD) (CDC = 6.53 ± 1.1 mm, CDD = 7.08 ± 1.2 mm) and the percentage of the carotid wall distension (%CWD = ((CDD-CDC)/CDC)*100%), which was %CWD = 7.67 ± 1.9. We found significant differences between the manual and the automated CDC and %CWD (($p = 0.13$ and $p = 0.92$) and no significant differences for the CDD ($p = 0.006$).

Manual Segmentation   Snakes base Segmentation

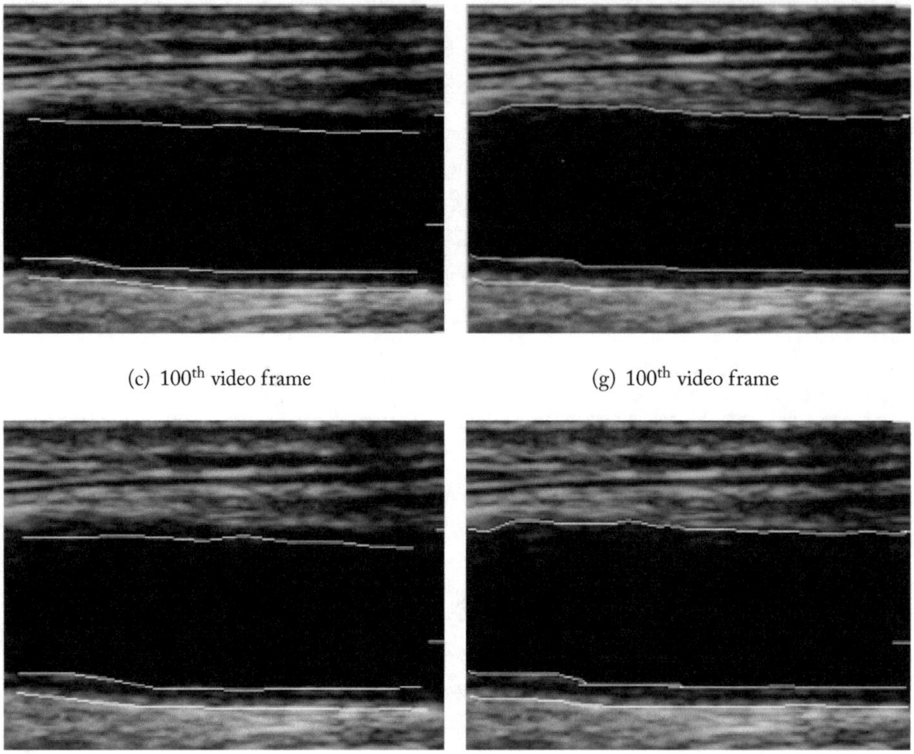

(c) 100th video frame   (g) 100th video frame

(d) 150th video frame   (h) 150th video frame

Figure 2.3: *(Continued.)* Manual (left column) and automated (right column) video segmentation of the CCA for the IMT and carotid diameter, at the (c) 100th (IMT_{man} = 0.92 mm, $diameter_{man}$ = 6.98 mm, IMT = 0.95 mm, $diameter$ = 7.21 mm), and (d) 150th (IMT_{man} = 0.90 mm, $diameter_{man}$ = 7.11 mm, IMT = 0.93 mm, $diameter$ = 7.31 mm) video frames of the video (video no. 6 in Table 2.6). Source [118], © IEEE 2012.

Table 2.6 presents the results of the evaluation metrics between the manual and the automated delineations of the IMT (average (±sd)) in millimetres, for all 10 videos of the CCA with RMSE = 0.43±0.14 mm, MSE = 0.20±0.11, MAE = 2.35±1.4, and MARE = 0.05±0.02, respectively.

The fully automated method for the video segmentation of the IMC presented in [118] can also be favourably applied in the clinical praxis so that the IMT changes over the cardiac cycle can be evaluated. A limitation of the method is that the IMC in the near wall of the CCA was not properly detected due to the nature of ultrasound imaging, thus the segmentation algorithm

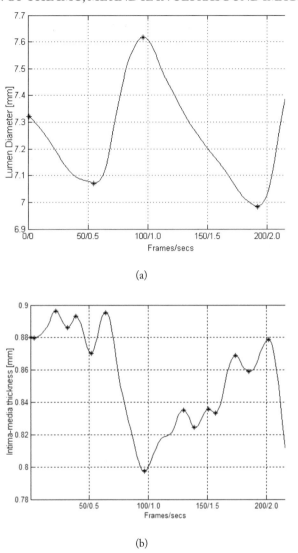

(a)

(b)

Figure 2.4: Lumen rate of change for the video of Fig. 2.3, showing in (a) the diameter change over two cardiac cycles, and in (b) the IMT rate of change for one cardiac cycle. Source [118], © IEEE 2012.

is developed for segmentation of the IMC at the far wall, where the borders are visible [119]. The success of the algorithm is due to the use of the available anatomical information regarding the position and general structure of the CCA to establish the best possible outcome. The algorithm was evaluated in [118], on videos from 10 cases with a mean IMT difference of 0.09 mm. The

Table 2.4: Mean, median, minimum and maximum values for the manual and their corresponding automated IMT detection in all 10 videos of the CCA. Standard deviations (sd) are given in parentheses.

	MANUAL	AUTOMATED			
Video	Mean [mm]	Mean [mm]	Median [mm]	Min [mm]	Max [mm]
1	0.62 (0.17)	0.64(0.12)	0.63(0.11)	0.61 (0.13)	0.67(0.09)
2	0.68(0.19)	0.7(0.11)	0.7(0.10)	0.63(0.10)	0.74(0.09)
3	0.73(0.22)	0.73(0.09)	0.73(0.09)	0.62(0.11)	0.78(0.10)
4	0.92(0.19)	0.96(0.10)	0.97(0.11)	0.92(0.10)	1.01(0.11)
5	1.11(0.16)	1.14(0.08)	1.14(0.11)	1.1(0.10)	1.19(0.10)
6	0.86(0.21)	0.87(0.12)	0.86(0.11)	0.84(0.10)	0.91(0.10)
7	0.65(0.17)	0.67(0.12)	0.67(0.10)	0.62(0.11)	0.7(0.11)
8	0.47(0.16)	0.48(0.11)	0.48(0.12)	0.45(0.11)	0.52(0.09)
9	0.60(0.17)	0.61(0.10)	0.6(0.09)	0.54(.09)	0.68(0.09)
10	0.40(0.18)	0.4(0.09)	0.39(0.09)	0.36(0.11)	0.44(0.10)
Mean (std)	0.70 (0.19)	0.72 (0.22)	0.71 (0.21)	0.67 (0.19)	0.76 (0.21)
Min	0.40	0.4	1.14	1.1	0.44
Max	1.11	1.14	0.39	0.36	1.19

Source [118], © IEEE 2012.

algorithm works well even on difficult cases, but further evaluation is required. Future work will involve assessment of the algorithm on a much larger dataset and discussion of how inter- and intra-observer variability affects the evaluation of the results.

2.1.5 AN OVERVIEW OF IMC IMAGE AND VIDEO SEGMENTATION TECHNIQUES

A number of IMC segmentation techniques for image and video were proposed in the literature as presented in Table 2.7. Most of the presented studies cover the segmentation of the IMC in ultrasound images of the CCA and very few studies presented cover ultrasound video [117, 118, 120–143].

2.2 SEGMENTATION OF THE ATHEROSCLEROTIC CAROTID PLAQUE IN ULTRASOUND IMAGING AND VIDEO

It is well documented that the risk of stroke increases with the severity of carotid stenosis and is reduced after carotid endarterectomy [110]. The degree of internal carotid stenosis is the only

Table 2.5: Carotid diameter during contraction (CDC), carotid diameter during distension (CDD), % of carotid wall distension (%CWD), mean and median diastolic measurements for the lumen diameter in 10 videos of the CCA. All measurements are in mm.

	MANUAL			AUTOMATED		
Video	CDC	CDD	%CWD	CDC	CDD	%CWD
1	5.77	5.99	3.67	5.86	6.05	3.14
2	4.91	5.21	5.76	4.99	5.35	6.73
3	7.36	7.93	7.19	7.47	8.06	7.32
4	7.00	7.51	6.79	7.06	7.64	7.59
5	8.92	9.57	6.79	8.91	9.69	8.05
6	7.21	7.61	5.26	7.04	7.7	8.57
7	6.10	6.67	8.55	6.28	6.73	6.67
8	5.64	6,32	10.76	5.72	6.34	9.78
9	5.61	6.41	12.48	5.83	6.47	9.89
10	6.21	6.8	8.68	6.14	6.75	9.03
Mean	6.47	7.0	7.59	6.53	7.08	7.68
(sd)	(1.16)	(1.21)	(2.62)	(1.11)	(1.21)	(2.01)
Min	4.91	5.21	3.67	4.99	5.35	3.14
Max	8.92	9.57	12.48	8.91	9.69	9.89

CDC: Carotid diameter during contraction, CDD: Carotid diameter during distension, %CWD: Carotid wall distension, sd: standard deviation. Source [118], © IEEE 2012.

well-established measurement that is used to assess the risk of stroke [111]. Indeed, it is the only criterion at present used to decide whether carotid endarterectomy is indicated or not [112]. In [9], we proposed and evaluated an integrated plaque segmentation system based on normalization, speckle reduction filtering, and snakes segmentation. Four different snake segmentation methods were investigated in [9], namely: (i) the Williams & Shah [114], (ii) the Balloon [116], (iii) the Lai & Chin [121], and (iv) the GVF [122]. These were applied on 80 plaque ultrasound images of the CCA. The comparison of the four different plaque snakes segmentation methods showed that the Lai & Chin segmentation method, gave slightly better results although these results were not statistically significant when compared with the other three snakes segmentation methods. It was also shown that the application of normalisation and speckle reduction filtering prior segmentation, improves both the manual and the automated plaque segmentation results. It was also shown that the proposed plaque image segmentation system cannot only reduce significantly the time required for the image analysis, but also it can reduce the subjectivity that accompanies manual delineations and measurements.

Table 2.6: Evaluation metrics between the manual and the automated IMT segmentations of the IMC in mm

Video	MAE	MARE	RMSE	NMSE	CIMA
1	1.89	0.024	0.297	0.088	13.54
2	0.54	0.013	0.135	0.018	13.37
3	3.17	0.075	0.503	0.253	20.31
4	3.15	0.050	0.370	0.150	26.07
5	4.41	0.070	0.499	0.249	39.00
6	4.68	0.082	0.534	0.285	23.56
7	1.62	0.037	0.368	0.135	15.68
8	1.21	0.067	0.596	0.356	10.38
9	1.62	0.037	0.368	0.135	13.67
10	1.21	0.067	0.597	0.356	9.09
Mean (sd)	2.35 (1.4)	0.05 (0.02)	0.43 (0.14)	0.20 (0.11)	18.47 (9.1)

MAE: Mean absolute error, MARE: Mean absolute relative error RMSE: Relative mean square error, NMSE: Normalized mean square error, CIMA: Carotid intima-media area, sd: standard deviation. Source [118], © IEEE 2012.

2.2.1   METHODOLOGY FOR THE SEGMENTATION OF PLAQUE IN ULTRASOUND IMAGING

For the segmentation of plaque a total of 80 B-mode and blood flow longitudinal ultrasound images of the CCA bifurcation were selected at random representing different types of atherosclerotic plaque formation with irregular geometry typically found in this blood vessel (see Database 2 in Appendix A.4). The images were acquired by the ATL HDI-3000 ultrasound scanner.

An expert manually delineated the plaque borders, between plaque and artery wall, and those borders between plaque and blood, on 80 longitudinal B-mode ultrasound images of the carotid artery, after image normalization [1, 9, 11], and speckle reduction filtering using the IDF software toolbox [2]. The procedure for carrying out the manual delineation process was established by a team of experts and was documented in the ACSRS project protocol [62]. The correctness of the work carried out by a single expert was monitored and verified by at least another expert. Usually the plaques are classified into type I to type V as documented in [9, 62]. In this work the plaques delineated were of type II, III and IV because it is easier to make a manual delineation since the fibrous cap, which is the border between blood and plaque, is more easily identified. If the plaque is of type I, borders are not visible well. Plaques of type V produce acoustic shadowing and the plaque is also not visible well.

Brightness adjustments of ultrasound images were carried out as documented in Section 2.1.1.

Table 2.7: An overview of IMC segmentation techniques in 2D and 3D ultrasound imaging and video of the CCA

Study	Segmentation Method	UIR	N	IMT$_{mean}$±sd [mm]	Performance metrics	IMT error*10^{-3} [mm]
	Ultrasound Imaging 2D					
Dwyer [133]	Edge detection	Yes	38	0.74±0.14	MAD	41±64
Wendelhag [134]	Dynamic Programming	Yes	100	0.92±0.25	MAD, CV%=15%	80±70
Liguori [135]	Pattern recognition and edge detection	Yes	30	0.93±0.33	MAD	20±31
Liang [136]	Dynamic Programming	Yes	50	0.93±0.25	MAD	42±20
Faita [137]	First order absolute moment edge detector	Yes	42	0.56±0.14	MAD	1±35
Golemati [130]	Canny edge detector and Hough Transform	No	10	0.61±0.05	CV%=7%	65+32
Gutierrez [138]	Multiresolution active contours	Yes	30	0.72±0.14	CV%=19%	-
Selzer [119]	Dynamic edge segmentation	Yes	24	0.78±0.17	MAD, CV%=4.03%	-
Loizou [98]	Snakes	Yes	100	0.68±0.12	MAD, HD, CV%=12.5%, BA, Histograms	50±25
Cheng [116]	Snakes	Yes	32	0.65±0.16	MSE	92±31.5
Chan [139]	Anisotropic Diffusion	Yes	40	0.82±0.19	-	-
Rocha [140]	Hybrid dynamic programming	Yes	47	0.62±0.16	MAD, BA	-
Delsanto [129]	Snakes	No	151	-	MAD	-
Molinari [141]	Integrated approach	No	200	0.71±0.16	MAD	63±49.1
Molinari [142]	Multiresolution edge detection	Yes	365	0.91±0.44	MAD	78±112
Loizou [99]	Snakes	Yes	100	0.67±0.12	MAD, HD, CV%=12.6%, BA, Histograms	8±20
	Ultrasound Video					
Illea [117]	Spatially continuous vascular model	No	40	0.60±0.10	MAD	0.007±0.176
Loizou [118]	Snakes	Yes	43	0.72±0.22	MAD, CV%=13%, BA	0.008±0.02
Destrempes [120]	Bayesian model	No	30	0.84±0.21	HD	78±40

IMC: Intima media complex, CCA: Common carotid artery, UIR: User interaction required, ROI: Region of interest, MAD: Mean absolute distance, HD: Hausdorf distance, MSE: Mean square error, sd: standard deviation, CV%: Coefficient of variation, BA: Bland Altman plots, N: Number of images. Reproduced from [118], © IEEE 2012.

For despeckle filtering the *DsFlsmv* linear scaling filter [2, 7, 9, 11], utilising the mean and the variance of a pixel neighbourhood was used for 4 itterations. This filter was introduced in [19], and evaluated on ultrasound imaging of the carotid giving best results in [7].

In most of the cases a plaque is visualised in a B-mode longitudinal ultrasound image and its size confirmed in transverse section using colour blood flow imaging. However, uniformly echolucent plaques are not obvious on B-mode, and colour flow imaging is needed. These echolucent plaques are seen as black filling defects. PW Doppler is used to measure velocity in order to grade the degree of stenosis and blood flow can be detected at a specific depth by selecting the time interval between the transmitted and received pulses. In [9] the blood flow image was used, in order to extract the initial snake contour for the plaque borders in the carotid artery. The plaque snakes contour initialization procedure, carried out using both the blood flow and the B-mode images is described in [9, 11].

Four different snake segmentation methods were investigated in this book, namely: (i) the Williams & Shah [114], (ii) the Balloon [127], (iii) the Lai & Chin [121], and (iv) the GVF [122]. The parameter values for the four different snakes segmentation methods, were the same in all experiments. For the Williams & Shah snake, the strength, tension and stiffness parameters were equal to $\alpha_s = 0.6$, $\beta_s = 0.4$ and $\gamma_s = 2$, respectively. For the Lai & Chin the regularisation parameter, λ_π, was varied as documented in [120]. For the GVF snake the elasticity, rigidity and the regularisation parameters were, $\alpha_{GVF} = 0.05$, $\beta_{GVF} = 0$, and $\mu_{GVF} = 0.2$, respectively [122].

ROC analysis [113] was used to assess the specificity and sensitivity of the four segmentation methods by the true-positive fraction, TPF, and false-positive fraction, FPF, [113]. The FPF, is calculated when the expert detects a plaque (when a plaque is present) and the computerized method identifies it as so, whereas the FPF, is calculated when the expert detects no plaque and the computerized method incorrectly detects that there is a plaque present. The TNF fraction is calculated when the expert identifies no plaque and the computerized method identifies it as so (absent), whereas the FNF is calculated when the expert identifies plaque presence and the computerized method incorrectly identifies plaque absence. Ratios of overlapping areas, can also be assessed by applying the similarity kappa index, KI, [235], and the overlap index [236] which can be found in [9].

The Wilcoxon matched-pairs signed rank sum test was used in order to detect if for each metric, TPF, TNF, FPF, FNF, KI, overlap index, Sp, P, and F, a significant difference or not exists between all the segmentation methods at $p < 0.05$. The test was applied on all 80 segmented plaques for all different segmentation methods.

2.2.2   METHODOLOGY FOR THE SEGMENTATION OF PLAQUE IN ULTRASOUND VIDEO

A total of 43 B-mode longitudinal ultrasound videos were acquired (see Appendix A.4, Database 4).

An expert neurologist with more than 25 years of clinical experience manually delineated the plaque borders, between plaque and artery wall, and the borders between plaque and blood, every 20 frames on 43 longitudinal B-mode ultrasound videos of the CCA, after image normalization and speckle reduction filtering [1, 13] using MATLAB® software developed by our group. The manual segmentations traced in the first frame could be transferred in the second frame and then could be readjusted by the expert accordingly. In total, 538 (10,750 frames / 20 = 538) ultrasound frames of the CCA were delineated. On average 13 frames per video were manually delineated by the expert. The procedure used for carrying out the manual delineation was the one established and documented in the asymptomatic carotid stenosis and risk of stroke (ACSRS) project protocol [62] for still images of the CCA. The correctness of the work carried out by a single expert was monitored and verified by at least another expert. In cases where several plaques were located in the CCA, comprising multiple components, each plaque was segmented independently so that each contour initialization represents a different plaque.

For speckle reduction, the filter *DsFlsmv*, introduced in [27], and evaluated in [7, 9], was applied to each consecutive frame prior the plaque segmentation. The filters of this type utilize first order statistics such as the variance and the mean of a pixel neighbourhood and may be described with a multiplicative noise model [7]–[27]. The moving window size for the despeckle filter DsFlsmv was 5×5 and the number of iterations applied to each video frame was two, where a complete description of the filter and its parameters can be found in [1, 7, 11, 27]. An example of the application of the DsFlsmv filter is shown in Fig. 2.6 after image normalisation.

Before running the video plaque snakes segmentation algorithm, two different plaque initialization procedures [13] were investigated for positioning the initial snake contour in the first frame of the video, as close to the area of interest (plaque borders) as possible. Prior to both initialization methods all the video frames were normalized and despeckled with the *DsFlsmv* filter. The normalized despeckled first video frame of a CCA video is shown in Fig. 2.6a. It is also assumed that the carotid artery is properly imaged in the video according to the standard clinical guidelines.

The Williams & Shah snake segmentation method [114] was used to deform the snake and segment the plaque borders in each video frame. The snake contour, $v(s)$, adapts itself by a dynamic process that minimises an energy function ($E_{snake}(v, s)$) and is defined in [1, 13, 114]. For the Williams & Shah snake, the strength, tension and stiffness parameters were equal to $\alpha_s = 0.6$, $\beta_s = 0.4$ and $\gamma_s = 2$, respectively.

The M-mode image (see Fig. 2.7d) can be generated in such a way that it crosses all plaque borders having maximum motion in opposite directions [13]. The procedure of the M-mode generation is described in [13].

Figure 2.7, presents the first video frame of the cardiac cycle from a B-mode ultrasound video of the left CCA, after normalization and despeckle filtering acquired from a male symptomatic subject at the age of 64 (having a stent on the right CCA and a stenosis of 40–50%). Figure 2.7b presents the automated snakes segmentations of the plaque boundaries at the far wall of the CCA and the lumen segmentation at the near wall. The segmented atherosclerotic plaque is shown in Fig. 2.7c, whereas Fig. 2.7d presents the despeckled M-mode image generated from the CCA video for the first 1,200 frames (12 seconds).

Perpendicular lines that cross the major axis of the plaque (from Fig. 2.7c), were placed automatically at the major axis quintiles (20%, 40%, 60%, and 80%). By scanning the intensity values along the straight perpendicular line selected by the user, the M-mode image is generated, by taking this line as the Y-axis, and each frame of the video as the X-axis (see Fig. 2.7d). Four M-mode images were generated for each of the corresponding four perpendicular lines. The manual delineations, as well as all other measurements were performed (by the same expert as documented earlier in this section and in [13]) using a system implemented in MATLAB® from our group. The M-mode images were converted to binary images, and morphological operators were applied to smooth the edges. Then, edge detection was applied on each M-mode image in order to derive the initial near and far wall boundaries, and the rate of change [13]. The snakes segmentation system [9] was also used to refine the derived snakes contours (see Fig. 2.7e) found on the M-mode image. From the derived contours at the far and near wall boundaries of the M-mode image, the diastolic and systolic diameters (plaque-lumen) of the carotid artery were calculated. This was done by finding corresponding maxima and minima (and vise versa) at the near and far wall boundaries of the M-mode image and then estimating their difference, which is the diameter rate of change (see Fig. 2.7f). The extracted contours for the rate of change were averaged to form the final state diagram of the video (see Fig. 2.7f showing final states of the video), where diastolic and systolic frames were estimated at the maxima and minima of the curve.

From the final state diagram of the video in Fig. 2.7f, a full cardiac cycle was selected by identifying the starting and ending frames of the cycle. The frames of the video corresponding to the identified cardiac cycle were then extracted. The segmentation algorithm was then applied on the frames representing a full cardiac cycle.

The video segmentation methods were evaluated using the true-positive fraction, TPF, and the true-negative fraction, TNF, corresponding to sensitivity and specificity [113]. Ratios of overlapping areas, can also be assessed by applying the similarity kappa index, KI, and the overlap index as used in [9, 13].

In order to further evaluate the state based identification algorithm, the metrics RMSE, NMSE, MAE, MARE, the carotid diameter during contraction (CDC), the carotid diameter during distension (CDD) and the percentage of the carotid wall distension (%CWD). Furtermore, the radial and longitudinal movements in these videos have been also investigated, where regions of interest (ROI's) were selected for each CCA video (see Fig. 2.7a) on the first normalised despeckled frame [237].

More specifically, three ROI's, namely the adventitia wall (AW) at the near wall, the plaque wall (PW) and the media-adventitia wall (MAW) were selected at the far wall of the CCA, as well as two ROIs, namely the maximum plaque border (Pmax) and the minimum plaque border (Pmin) were selected (see also Fig. 2.7a). The radial (RS), longitudinal (LS) and shear strains (SS) from the radial and longitudinal displacement indices as documented in [13, 237] were also computed.

2.2.3   SEGMENTATION OF THE PLAQUE IN ULTRASOUND IMAGING

Figure 2.5 illustrates an original longitudinal ultrasound B-mode image of a carotid plaque with a manual delineation made by the expert in (a), and the results of the William & Shah segmentation in (b), the Balloon segmentation in (c), the Lai & Chin segmentation in (d), and the GVF segmentation in (e). Figure 2.5f shows the superimposition of the segmentation contours computed in Fig. 2.5b–e. As illustrated in Fig. 2.5f, both the manual and the snakes segmentation contours are visually very similar.

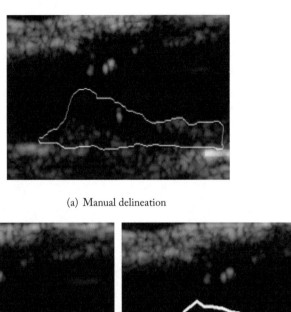

(a) Manual delineation

(b) Williams & Shah   (c) Ballon

Figure 2.5: Plaque segmentation results on a longitudinal ultrasound B-mode image of the carotid artery: (a) manual, (b) Williams & Shah, (c) Balloon Source [9], ©IEEE 2007. *(Continues.)*

It should be noted that the despeckle filter *DsFlsmv*, with a moving sliding window of 5×5, was iteratively applied for 4 itterations on all images prior to segmentation.

Table 2.8 presents a comparison of the four different plaque snakes segmentation methods (Williams & Shah, Balloon, Lai & Chin, and GVF) with the manual segmentation as performed by an expert on 80 longitudinal ultrasound images of the carotid plaque as described in [9]. Although all methods demonstrated similar performance, the best overall performance was demonstrated by the Lai & Chin snakes segmentation method. The results showed that the Lai & Chin snakes segmentation method, agrees with the expert in 80.89% of the cases, true negative fraction (TNF), by correctly detecting no plaque, in 82.70% of the cases, true positive fraction (TPF), by correctly detecting a plaque, disagrees with the expert in 15.59% of the cases, false negative frac-

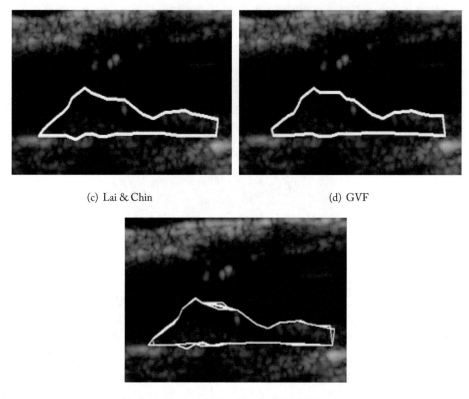

(c) Lai & Chin   (d) GVF

(e) Superimposition of segmentation contours computed in (b)–(e)

Figure 2.5: *(Continued.)* Plaque segmentation results on a longitudinal ultrasound B-mode image of the carotid artery: (d) Lai & Chin, (e) GVF snake, and (f) superimposition of segmentation contours computed in (b)–(e). Source [9], ©IEEE 2007.

tion (FNF), by detecting no plaque, and in 5.86% of the cases, false positive fraction (FPF), by detecting a plaque. The similarity kappa index, KI, and the overlap index, for the Lai & Chin snakes segmentation method were the highest, equal to 80.66% and 69.3%, respectively.

The best FPF, and FNF, fractions were given by the Balloon snakes segmentation method, with 5.4% and 13.90%, respectively. The GVF snakes segmentation method, showed for this experiment the worst results with the lowest similarity kappa index, KI, (77.25%), and the lowest overlap index (66.6%).

Table 2.9 presents a comparison of four different plaque snakes segmentation methods (Williams & Shah, Balloon, Lai & Chin, and GVF), on 80 longitudinal ultrasound images of the carotid plaque, based on the sensitivity, R, specificity, Sp, precision, P, and the measure F,

Table 2.8: ROC analysis for the four different plaque segmentation methods and the manual delineations made by an expert on 80 ultrasound images of the carotid artery

Segmentation Method	System Detects	Expert Detects no plaque	Expert Detects plaque	KI	Overlap Index
Williams& Shah	No plaque	TNF=77.59%	FNF=19.64%	78.86 %	67.60 %
	Plaque	FPF=6.50%	TPF=81.76%		
Balloon	No plaque	TNF=77.12%	**FNF=13.90%**	77.87 %	67.79 %
	Plaque	**FPF=5.40%**	TPF=80.35%		
Lai & Chin	No plaque	**TNF=80.89%**	FNF=15.59%	**80.66 %**	**69.30 %**
	Plaque	FPF=5.86%	**TPF=82.70%**		
GVF	No plaque	TNF=79.44%	FNF=14.90%	77.25 %	66.60 %
	Plaque	FPF=6.30%	TPF=79.57%		

FPF: False positive fraction, FNF: False negative fraction, TPF: True positive fraction, TNF: True negative fraction,

KI: Williams index. Source [9], © IEEE 2007

as described in [46]. Bolded values in Table 2.4 show best performance of the segmentation algorithms. The best sensitivity, R, was given by the Lai & Chin (0.827), followed by the Williams & Shah (0.8176), whereas the best specificity, Sp, was given by the Balloon (0.9460), followed by the Lai & Chin (0.9416) snakes segmentation method. The Lai & Chin gave the best precision, P, (0.9338), which is better than the rest of the segmentation methods, whereas the best F, was given by the Balloon (0.8882), followed by the Lai & Chin (0.8851) snakes segmentation method.

Table 2.9: ROC analysis for the four different plaque segmentation methods and the manual delineations made by an expert on 80 ultrasound images of the carotid artery based on the sensitivity, R, specificity, Sp, precision, P, and 1-Effectiveness measure, 1-E

Segmentation Method	Sensitivity (R)	Specificity (Sp)	Precision (P)	F=1-E
Williams & Shah	0.8176	0.9350	0.9263	0.8621
Balloon	0.8053	**0.9460**	0.9271	**0.8882**
Lai & Chin	**0.8270**	0.9416	**0.9338**	0.8851
GVF	0.7957	0.9370	0.9266	0.8824

Source [9], © IEEE 2007.

2.2.4 RESULTS OF THE SEGMENTATION OF PLAQUE IN ULTRASOUND VIDEO

The predictive ability to identify which patients will have a stroke is poor, where the current practice of assessing the risk of stroke relies on measuring the thickness of the CCA wall (Intima-Media-Thickness, IMT) [136, 143] or the artery lumen stenosis by identifying the plaque borders in the carotid artery [9]. While the IMT and the degree of stenosis can reliably be delineated in still B-mode ultrasound images [13, 98, 108, 144, 145] the moving borders of the atherosclerotic carotid plaque and the identification of the systolic and diastolic video states [120], may provide additional information of an individual's stroke risk, and treatment of asymptomatic patients may be improved [146].

We proposed in [13] an integrated system for the video segmentation of the CCA based on our previous work [9, 143] in order to further facilitate the quantitative assessment of atherosclerosis disease. The main extension to our previous work in [9], which was based on still image segmentation, was the consideration of moving frames, the identification of the states diagram of the video, the extraction of systolic and diastolic states [120], and the automated re-initialisation of the snake contour every 20 frames based on recent work presented in [143].

To the best of our knowledge there is only one published study for the segmentation of atherosclerotic carotid plaque in ultrasound CCA videos [120]. The method by Destrempes et al. [120] is based on a Bayesian segmentation model and is evaluated on 33 video sequences. Still, several other studies investigated the segmentation of atherosclerotic carotid plaque in ultrasound images. An overview of these techniques is given in Table 2.14. In [13], we introduced an integrated system for the segmentation of the atherosclerotic carotid plaque in 2D ultrasound video of the common carotid artery (CCA) is presented and evaluated. The system builds some of the authors' previous work [9], and incorporates image frame normalization, speckle reduction filtering, initial contour estimation, M-mode generation and snakes segmentation for the advancement of evaluation and treatment of the carotid atherosclerosis.

We present in Fig. 2.6 the 100th frame of a symptomatic video for the original and the despeckled frames with filters *DsFlsmv*, *DsFhmedian*, *DsFkuwahara*, and *DsFsrad* when applied to the whole frame (left column) and to an ROI selected by the user of the system (right column), respectively. The automated plaque segmentations performed by the integrated system proposed in [13] are also shown in the images. The filters *DsFlsmv* and *DsFhmedian* smoothed the video frame without destroying subtle details.

Figure 2.7 and Fig. 2.8 presents the application of the *DsFlsmv* (see Fig. 2.7a, c, and e) and *DsFhmedian* (see Fig. 2.8b, d, and f) despeckle filters, which showed best performance (see also Table 3.1–Table 3.3), on consecutive video frames (1, 50 and 100) of a symptomatic subject, for the cases where the filtering was applied on the whole video frame (see left columns of Fig. 2.7 and Fig. 2.8) and on an ROI selected by the user (see right columns of Fig. 2.7 and Fig. 2.8).

Figure 2.9 presents two different examples of video plaque segmentation with atherosclerotic plaques appearing at the CCA in the left and right column respectively. The segmentations were performed for the first and second cardiac cycle, for the 50th, 60th, 70th, and 150th video frames of the video, respectively. In the left column a case from a female asymptomatic subject is presented with three different plaques at the near and far wall of the CCA, at the age of 61, with a left CCA stenosis of 35–40%. In the right column one can see a case from a male symptomatic subject at the age of 64, a stent on the right CCA and with a stenosis of 40–50%. It is observed that the proposed segmentation algorithm is able to follow the plaque borders consistently as a result of the initialisation procedures followed.

Despeckle filtering on the whole frame   Despeckle filtering on the ROI

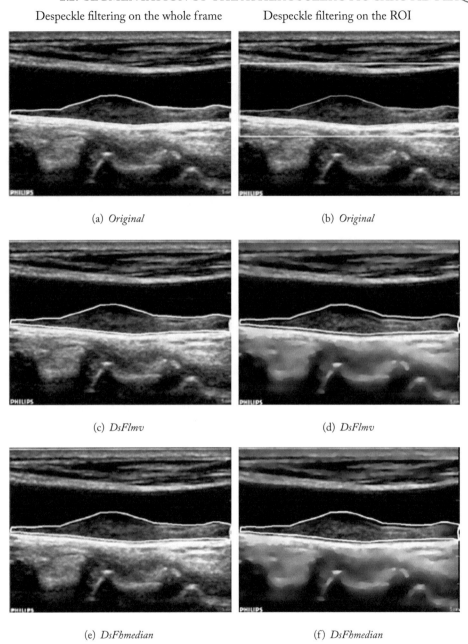

(a) *Original*   (b) *Original*

(c) *DsFlmv*   (d) *DsFlmv*

(e) *DsFhmedian*   (f) *DsFhmedian*

Figure 2.6: Examples of despeckle filtering on a video frame of a symptomatic CCA video for the whole image frame in the left column, and on an ROI (including the plaque (shown in (b)), in the right column for: (a), (b) original, (c), (d) *DsFlsmv*, (e), (f) *DsFhmedian*. The automated plaque segmentations are shown in all examples. Source [14], © IEEE 2012. *(Continues.)*

Despeckle filtering on the whole frame   Despeckle filtering on the ROI

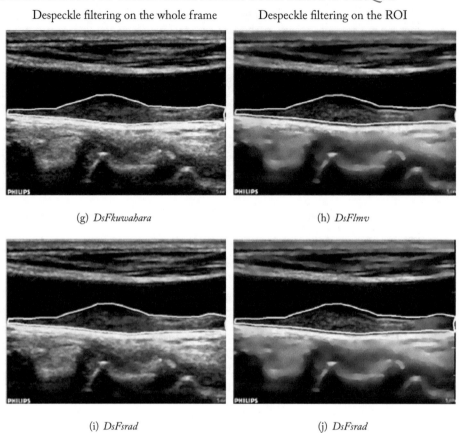

(g) *DsFkuwahara*   (h) *DsFlmv*

(i) *DsFsrad*   (j) *DsFsrad*

Figure 2.6: *(Continued.)* Examples of despeckle filtering on a video frame of a symptomatic CCA video for the whole image frame in the left column, and on an ROI in the right column for: (g), (h) *DsFkuwahara*, and (i), (j) *DsFsrad*. The automated plaque segmentations are shown in all examples. Source [14], © IEEE 2012.

Table 2.10 tabulates the quantitative results of the statistical analysis based on TNF, TPF, KI and overlap index for the proposed video segmentation method performed for 1–2 cardiac cycles, on 43 ultrasound videos of the CCA for the two different initialisation methods (see two last rows in Table 2.6). The results of the automated segmentation method are compared with the manual tracings of the expert which are considered to be the ground truth. The results show that the proposed method using the two different initialisation methods, (mean ±sd method 1 / mean ±sd method 2), agrees with the expert by correctly detecting no plaque (TNF) in (83.7±7.6% / 84.3±7.5%) of the cases, by correctly detecting a plaque (TPF) in (85.42±8.1% / 86.1±8.0%) of the cases. The similarity kappa index KI, and the overlap index, for the proposed video snakes

Despeckle filtering on the whole frame   Despeckle filtering on the ROI

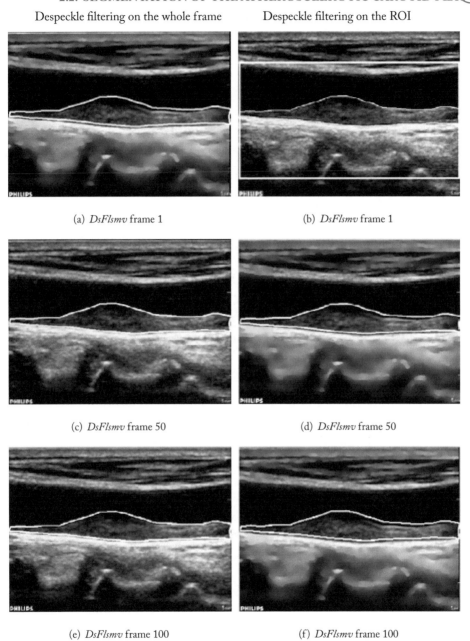

(a) *DsFlsmv* frame 1   (b) *DsFlsmv* frame 1

(c) *DsFlsmv* frame 50   (d) *DsFlsmv* frame 50

(e) *DsFlsmv* frame 100   (f) *DsFlsmv* frame 100

Figure 2.7: Examples of despeckle filtering with the filter DsFlsmv ((a)–(f)) on a symptomatic video of the CCA with plaque at the far wall of the CCA on frames 1, 50 and 100 for the whole image frame in the left column, and on an ROI, in the right column. The automated plaque segmentations are shown in all examples. Source [14], © IEEE 2012.

Despeckle filtering on the whole frame   Despeckle filtering on the ROI

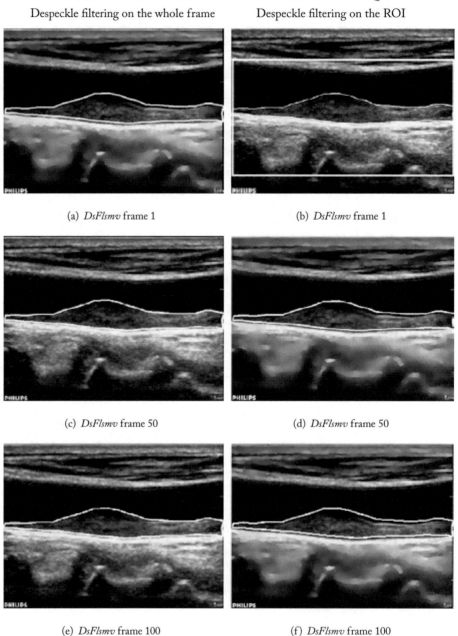

(a) *DsFlsmv* frame 1   (b) *DsFlsmv* frame 1

(c) *DsFlsmv* frame 50   (d) *DsFlsmv* frame 50

(e) *DsFlsmv* frame 100   (f) *DsFlsmv* frame 100

Figure 2.8: Examples of despeckle filtering with the filter *DsFhmedian* ((a)–(f)) on a symptomatic video of the CCA with plaque at the far wall of the CCA on frames 1, 50 and 100 for the whole image frame in the left column, and on an ROI in the right column. The automated plaque segmentations are shown in all examples. Source [14], © IEEE 2012.

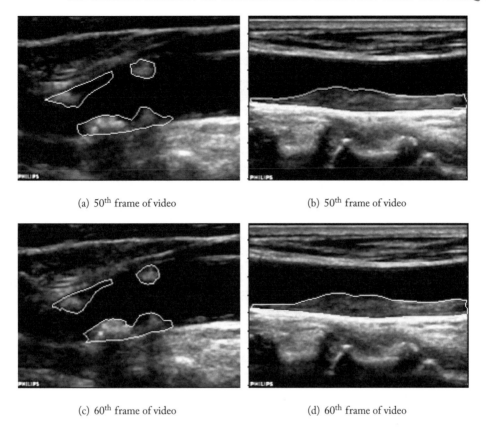

(a) 50th frame of video   (b) 50th frame of video

(c) 60th frame of video   (d) 60th frame of video

Figure 2.9: Two different examples of plaque segmentation of plaques appearing at the CCA in the left and right column, respectively, at the 50th and 60th video frames of the videos. Source [13], © IEEE TUFFC. *(Continues.)*

segmentation method were, equal to 84.6% / 85.3% and 74.7% / 75.4% using the two different initialisation techniques, respectively. There was no significant difference between the two methods for all metrics investigated using a pair t-test at $p < 0.05$. However, a small improvement was observed in almost all evaluation metrics when the segmentations were performed with the second initialisation method. It should further be noted that the snake contour may be attracted occasionally to local minima and converge to a wrong location. This occurred in less than 10% of the cases (i.e., in 4 videos).

Figure 2.10 presents the M-mode image of the video, where the video states are superimposed at the far and near wall respectively. The insert on the left upper side of the images in Fig. 2.10d and Fig. 2.10e, indicates the straight perpendicular line that can also be selected by the user of the proposed system in order to generate the M-mode image. Finally, in Fig. 2.10f

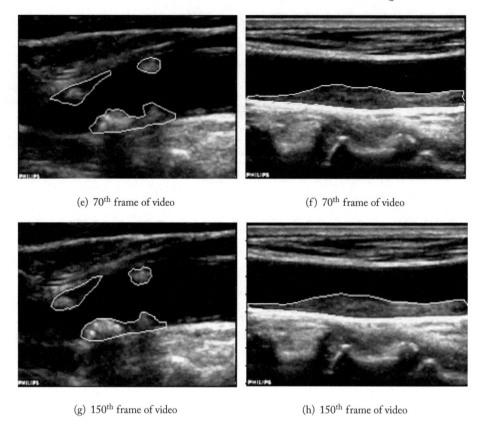

(e) 70th frame of video   (f) 70th frame of video

(g) 150th frame of video   (h) 150th frame of video

Figure 2.9: *(Continued.)* Two different examples of plaque segmentation of plaques appearing at the CCA in the left and right column, respectively, at the 70th and 150th video frames of the videos. Source [13], © IEEE TUFFC.

the diameter change is presented with the help of a step diagram showing systolic (1, 36, 168, 240, 372, 464, 543, 629, 728, 803, 896, 993) and diastolic frames (49, 106, 208, 306, 400, 483, 592, 680, 747, 844, 931) of the video. The maximum carotid diameter during distension (2 mm at frame 543 indicated with an *) and maximum carotid diameter during contraction (2.67 mm at frame 844 indicated with an *) are also shown.

Table 2.11 presents the results of the evaluation metrics between the manual and the automated state diagrams (mean ±sd) in microseconds for the asymptomatic and the symptomatic subjects respectively. The RMSE, NRMSE, MAE and the MARE were for all videos (125.41±11.77) μsecs, (121.16±10.38) μsecs, (119.06±11.96) μsecs and (5.43±0.95) μsecs respectively. The t-test test gave statistically significant differences between the asymptomatic and symptomatic cases for all evaluation metrics tabulated in Table 2.11.

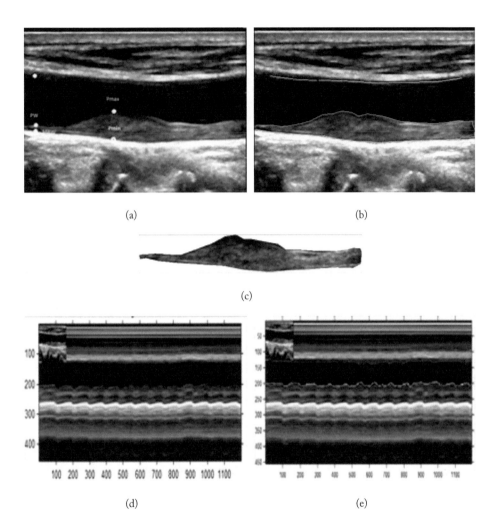

Figure 2.10: Illustration of the M-mode generation (see section II.G for the implementation details), (a) first normalized and despeckled (with *DsFlsmv*) frame of a B-mode ultrasound video of the CA, (b) segmentation of the plaque boundaries and the near wall of the CA by snakes, (c) extracted plaque, (d) despeckled M-mode image generated from the CA video for a selected B-mode line, e) initial M-mode states superimposed on the original M-mode image at the far and near walls, respectively. Source [13], © IEEE TUFFC. *(Continues.)*

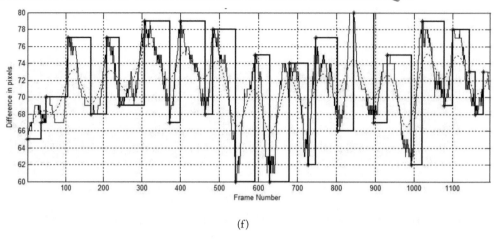

(f)

Figure 2.10: *(Continued.)* Illustration of the M-mode generation (see section II.G for the implementation details), (f) diameter change (averaged across the major plaque axis quintiles, with step diagram and systolic and diastolic frames of the video with maximum carotid diameter during distension (*) and maximum carotid diameter during contraction. Diastolic and systolic frames (from 0-1200) (100 frames per second= 12 seconds). Contraction frames: 1, 36, 168, 240, 372, 464, 543, 629, 728, 803, 896, 993. Distension Frames: 49, 106, 208, 306, 400, 483, 592, 680, 747, 844, 931. Minimum Carotid Diameter: 2mm at frame 543. Maximum Carotid Diameter: 2.67 mm at frame 844. In Fig. (a) we illustrate the points adventitia wall (AW), plaque wall (PW), media-adventitia wall (MAW), plaque at maximum point (P_{max}) and plaque at minimum point (P_{min}) (see [130] for definition). Source [13], © IEEE TUFFC.

Table 2.12 illustrates the results of the CDC, CDD and the %CWD for the asymptomatic and the symptomatic subjects (mean±sd). The CDC, CDD and the %CWD were for the asymptomatic/the symptomatic groups 5.13±0.34 / 5.48±0.29 mm, 5.89±0.31/6.27±0.35 mm, 10.2±1.03 / 16.09±0.83%, respectively. The t-test, shown in the last row of Table 2.12, performed between the values obtained for the asymptomatic and symptomatic cases, at $p < 0.05$, gave no statistically significant differences between them.

Table 2.13 presents the strain indices (mean±sd) over two consecutive cardiac cycles for the asymptomatic and symptomatic groups. More specifically the following measures were evaluated, a radial strain at wall (RSW) of (5.13±0.34)%, a longitudinal strain of (2.84±0.65)%, a shear strain at wall (SSW) of (0.42±0.12) rad, a shear strain at plaque (SSP) of (0.06±0.01) rad, and a radial strain at plaque (RSP) of (3.13±1.21) rad, respectively. All strain indices between the asymptomatic and symptomatic subjects were found to be not significantly different.

Table 2.10: Performance metrics of sensitivity (TPF), specificity (TNF), KI, and overlap index for the video segmentation methods performed on 43 ultrasound videos of the CCA for the two different initialization methods. Manual segmentations performed by an expert were used

Segmentation Method	Ultrasound Data	Sensitivity (TPF)	Specificity (TNF)	KI	Overlap Index
Loizou [46]	Images (N=80)	82.7%	80.9%	80.7 %	69.3 %
Golemati [129]	Images (N=4)	97.5±1.0%	96±10%	-	-
Destrempes [119]	Videos (N=33)	83.7±8.3%	94.1±4.2%	0.85±0.75	0.75±0.1
Loizou [50] (intensity based initialization)	Videos (N=43)	85.4 ±8.1%	83.7±7.6%	84.6%	74.7%
Loizou [50] (initial contour based initialization)	Videos (N=43)	86.1±8.0%	84.3±7.5%	85.3%	75.4%

TNF: True-negative fraction, TPF: true-positive fraction, KI: Kappa index. Source © IEEE TUFFC [13]

Table 2.11: Mean±sd evaluation metrics between the manual and the automatic states in μsecs

Video	RMSE	NRMSE	MAE	MAR
Asymptomatic (N=38)	191.25±13.31	212.79±12.42	182.63±19.76	7.12±1.01
Symptomatic (N=5)	59.56±10.23	43.53±8.34	55.50±4.20	3.74± 0.89
t-test between Asymptomatic and Symptomatic[1]	S (p=0.04)	S (p=0.002)	S (p=0.031)	S (p=0.029)

RMSE: Relative mean square error, NRMSE: Normalized mean square error, MAE: Mean Absolute error, MARE: Mean absolute relative error. [1]Test carried out at p<0.05. S: Significantly different. Source © IEEE TUFFC [13]

Table 2.12: Mean ±sd carotid diameter during contraction (CDC) in [mm], carotid diameter during distension (CDD) in [mm] and % of carotid wall distension (%CWD) for the automated segmentations

Video	CDC [mm]	CDD [mm]	%CWD
Asymptomatic (N=38)	5.13±0.34	5.89±0.31	10.2±1.03
Symptomatic (N=5)	5.48±0.29	6.27±0.35	16.09±0.83
t-test between Asymptomatic and Symptomatic[1]	NS (p=0.59)	NS (p=0.11)	NS (p=0.38)

[1]Test carried out at p<0.05. NS: non-significantly different. Source © IEEE TUFFC [13]

2.2.5 AN OVERVIEW OF PLAQUE SEGMENTATION TECHNIQUES

An overview of different plaque segmentation techniques for ultrasound images and videos are given in Table 2.14. Important findings and additional details are also given for each method. Hamou et al. [123], proposed a method based on the Canny edge detector to detect the plaque in longitudinal CCA ultrasound images. A morphological based approach for the carotid contour extraction was proposed in [124] for longitudinal ultrasound images of the CCA, incorporating speckle reduction filtering, contour quantization, morphological contour detection, and a contour enhancement stage. Mao et al. [125], proposed a discrete dynamic contour model for extracting the carotid artery lumen in 2D transversal ultrasound images, whereas Abolmaesumi et al. [126], introduced a method based on the star algorithm improved by Kalman filtering, for the plaque segmentation in transversal carotid ultrasound images. A semi-automatic method for plaque segmentation in 3D images of the CCA using the Balloon model introduced in [127] was proposed by Gill et al. [128]. Loizou et al. [9], proposed a plaque segmentation method, which was based on the Williams & Shah snake [114], for the extraction of the CCA using an

Table 2.13: Mean±sd values of strain indices of the carotid artery for asymptomatic (N=38) and symptomatic (N=5) subjects over two consecutive cardiac cycles

Strain Indices	Asymptomatic (n=38)	Symptomatic (n=5)	t-test[1]
RSW (%)	5.13±0.34	5.37±0.26	NS (0.31)
LS (%)	2.84±0.65	2.97±0.42	NS (0.72)
SSW (rad)	0.42±0.12	0.49±0.17	NS (0.43)
SSP (rad)	0.06±0.01	0.12±0.03	NS (0.096)
RSP (rad)	3.13±1.21	3.9±0.98	NS (0.59)

RSW: Radial strain at wall, LS: Longitudinal strain at wall, SSW: Shear strain at wall, SSP: Shear strain at plaque, RSP: Radial strain at plaque.
[1]Test carried out at p<0.05, NS: Non-Significantly different. Source © IEEE TUFFC [13]

automated contour estimation and applied on 80 ultrasound images of the CCA. The atherosclerotic carotid plaque in [129], was segmented in 56 2D longitudinal ultrasound images using a gradient based snake segmentation method and fuzzy K-means algorithm with an initialization based on pixel intensity. In [130], the Hough transform was applied to perform segmentation of plaque in 4 2D cross sectional ultrasound images of the CCA. In [131] an automated segmentation method for 3D ultrasound carotid plaque based on a geometrically deformable model, taking advantage of both the local and regional detectors, was proposed. More recently in [132], a 3D semi-automated segmentation method using sparse field level sets, where the users choose anchor points on each boundary, was proposed for the segmentation of 3D CCA images. All the above methods, with the exception of [120] (where ultrasound video segmentation of the plaque in the CCA was proposed), have investigated and proposed different solutions for the segmentation of the atherosclerotic carotid plaque in ultrasound images of the CCA.

As shown in Table 2.14, different methods were investigated for the segmentation of the atherosclerotic carotid plaque in ultrasound images, yet these studies were evaluated on a limited number of subjects. Therefore, the need still exists for the development, implementation, and evaluation, of an integrated system enabling the automated segmentation of ultrasound imaging carotid plaque.

2.3   DISCUSSION ON DESPECKLING OF THE INTIMA MEDIA COMPLEX AND THE PLAQUE IN IMAGING AND VIDEO

It was shown in section 2.1.2 that normalization and despeckle filtering improves the outcome of the IMT segmentation and produces more accurate and reproducible results when compared with the manual segmentation method. Speckle reduction filtering of the carotid artery was also applied in [51], where it was also shown that this improves the image quality and visual interpretation of the experts. Furthermore, in [7], it was shown that image normalization followed by speckle reduction filtering produces better quality images, whereas the reverse (speckle reduction

Table 2.14: An overview of atherosclerotic carotid plaque segmentation techniques in ultrasound imaging and video

Study	Segmentation Method	AIC	UI	Performance	TPF (%)	N
Ultrasound Imaging 2D						
Hamou [125]	Canny edge detection	No	-	-	-	-
Abdel-Dayen [126]	Morphological based	No	-	-	-	-
Mao [127]	Discrete dynamic contour	No	No	-	-	7
Abolmaesumi [128]	Kalman filtering	No	-	-	-	1
Gill [130]	Balloon	No	No	-	-	2
Loizou [9]	Active contour model	Yes	Yes	KI=80.66%, O=66.6%, Sp=0.937, P=0.926	(82.70±2.1)	80
Delsanto [131]	Gradient based snake with fuzzy k-means	Yes	No	-	-	56
Golemati [132]	Hough transform	No	-	-	(97.5±1.0)	4
Ultrasound Video 2D						
Destrempes [122]	Bayesian model	No	Yes	TNF=(83.7±8.3)%, KI=84.8%, O=74.6%, HD=(1.24±0.4)mm, MPD=(0.24±0.08)mm	(83.7±8.3)%	33
Loizou [13]	Snakes	Yes	Yes	TNF=(84.3±7.5)%, KI=85.3%, O=75.4%	(86.1±8.0)%	43
Ultrasound Imaging 3D						
Zahalka [133]	Geometrically deformable model	Yes	Yes	-	TPF=95	1
Ukwatta [134]	Level sets	No	Yes	CV%=5.1%, ISD=0.2±0.1mm	94.4±2.2	21

AIC: Automatic initial contour, UI: User interaction, TPF: True-positive fraction, TNF: True-negative fraction, KI: Williams's index,
O: Overlap, Sp Specificity, P: Precision, HD: Hausdorff distance, MPD: Mean point distance, CV%: Coefficient of variation,
N: Number of cases investigated. Source [13] © IEEE 2014

filtering followed by normalization) might produce distorted edges. The preferred method is to apply first normalization and then speckle reduction filtering for better results. Speckle reduction filtering of the carotid was also proposed by [7–9] and [98], where it was shown that this improves the image quality and the visual evaluation of the image.

The comparison of the four different plaque snakes segmentation methods, proposed in our recent study [9] and also presented in Section 2.2.2, for the segmentation of the atherosclerotic carotid plaque from ultrasound images, showed that the Lai & Chin snakes segmentation method, gave slightly better results although these results were not statistically significant when compared with the other three snakes segmentation methods (Williams & Shah, Balloon, and GVF). To the best of our knowledge no other study carried out ultrasound image normalization as described in this study, prior to the segmentation of the atherosclerotic carotid plaque. However, in [123], histogram equalization was performed on carotid artery ultrasound images for increasing the image contrast. The normalization method proposed in this book was documented to be helpful in the manual contour extraction as well as in the snake's segmentation of the IMT [8, 98] and plaque [9]. Moreover, this method increased the classification accuracy of different plaque types as assessed by the experts [34, 112]. It was also shown in [148], that speckle reduction increases the percentage of the correct characterization the carotid artery atherosclerotic plaques in ultrasound imaging. It was also shown that the image quality was improved because of better structural differentiation due to suppression of reverberation artifacts. The study in [148] concluded that routine use of speckle reduction may contribute to better diagnostic confidence in the carotid artery territory, including more precise visualization of plaque morphological details. Furthermore, in another study [149], a review of speckle tissue characterization methods based on statistical, model-based, signal processing, and geometrical approaches was presented. It was documented that these methods are often applied to confirm the speckle nature of the elements. Characterizing speckle in ultrasound images leads, in turn, to tissue characterization which is a very important issue in ultrasound medical imaging. Therefore, speckle detection comes as an initial step before ultrasound image segmentation, registration, motion tracking, and various other quantitative tasks for clinical practice and decision making.

CHAPTER 3

Evaluation of Despeckle Filtering of Carotid Plaque Imaging and Video Based on Texture Analysis

In this chapter we present the methods of texture analysis, image quality evaluation metrics, distance measures, univariate statistical analysis and the kNN classifier, which are used to evaluate despeckle filtering (see also companion volume I [1]) on imaging and video. Finally, the procedure of visual despeckle filtering evaluation carried out by experts is presented.

For speckle reduction, the 16 different despeckle filtering methods, already described in the companion monograph (Volume I) [1], were applied to each image or video prior to IMC or plaque segmentation. Despeckle filtering was applied after image or video normalisation (see [1]), either to the entire image or to an ROI, (see also Fig. 3.1) selected by the user. The selected ROI can be of any shape but the IDF software [2] doesn't support multiple ROIs selection. In the latter case, where the user of the system is interested only in the selected ROI, the area outside the ROI can be blurred using the *DsFlsmv* filter operating with a sliding moving window of [15 × 15] pixels and a number iterations 5 (see also Fig. 3.1 and Fig. 3.2). It should be noted that the blurring is applied outside of the ROI if the user of the system is not interested to subjectively evaluate this area. The input parameters of the 16 different despeckle filters for the IDF and VDF software toolboxes can be selected by the user as it was documented in [1–3]. The 16 despeckle filters evaluated in this chapters, were applied on 220 asymptomatic and 220 symptomatic ultrasound images of the CCA. Four despeckle filters (*DsFlsmv*, *DsFhmedian*, *DsFkuwahara*, *DsFsrad*) were further applied to 10 videos of the carotid artery bifurcation. A total of 61 texture features [1], were extracted from the original and despeckle images and videos and the most discriminant ones are presented. The performance of these filters is investigated for discriminating between asymptomatic and symptomatic images using the statistical kNN classifier. Moreover, 16 different image quality evaluation metrics [1] (see also Section A.4) were computed, as well as visual evaluation scores carried out by two experts.

Additionally, we present the image quality evaluation results performed on 80 ultrasound images of the CCA, which were acquired from two different ultrasound imaging scanners (ATL

HDI-3000 and ATL HDI-5000). Textures features and image quality evaluation metrics were also extracted from the original and the despeckled images from both scanners and two experts evaluated the images visually. We will further consider, in this chapter, the problem of filtering multiplicative noise in ultrasound videos of the CCA in order to increase the visual interpretation by experts and facilitate the automated analysis of the videos. We will apply and demonstrate the video despeckling techniques by investigating their performance on 10 ultrasound videos of the CCA. The despeckling filtering techniques were evaluated through visual perception evaluation, performed by two medical experts as well as through a number of texture characteristics and video quality metrics. These were extracted from the original and the despeckled videos.

3.1   EVALUATION OF DESPECKLE FILTERING ON CAROTID PLAQUE IMAGING BASED ON TEXTURE ANALYSIS

Following the despeckling, texture features may be extracted from the original and the despeckled images and videos using the IDF [2] and VDF [3] toolboxes, respectively. Texture image and video analysis is one of the most important features used in image processing and pattern recognition. It can provide information about the arrangement and spatial properties of fundamental image elements. Texture provides useful information for the characterization of atherosclerotic plaque [33]. In this study a total of 65 different texture features were extracted both from the original and the despeckled ultrasound images and videos as follows [33, 152]:

1. *Statistical Features (SF)*: 1) Mean, 2) Median, 3) Variance (σ^2), 4) Skewness (σ^3), 5) Kurtosis (σ^4), and 6) Speckle index (σ^2/m).

2. *Spatial Gray Level Dependence Matrices (SGLDM)* as proposed by Haralick et al. [152]: 1) Angular second moment, 2) Contrast, 3) Correlation, 4) Sum of squares: variance, 5) Inverse difference moment, 6) Sum average, 7) Sum variance, 8) Sum entropy, 9) Entropy, 10) Difference variance, 11) Difference entropy, 12), and 13) Information measures of correlation. Each feature was computed using a distance of one pixel. Also for each feature the mean values and the range of values were computed, and were used as two different features sets.

3. *Gray Level Difference Statistics (GLDS)* [153]: 1) Contrast, 2) Angular second moment, 3) Entropy, and 4) Mean.

4. *Neighbourhood Gray Tone Difference Matrix (NGTDM)* [29]: 1) Coarseness, 2) Contrast, 3) Business, 4) Complexity, and 5) Strength.

5. *Statistical Feature Matrix (SFM)* [30]: 1) Coarseness, 2) Contrast, 3) Periodicity, and 4) Roughness.

6. *Laws Texture Energy Measures (TEM)* [30]: For the laws TEM extraction, vectors of length $l = 7$, $L = (1, 6, 15, 20, 15, 6, 1)$, $E = (-1, -4, -5, 0, 5, 4, 1)$ and $S = (-1, -2, 1, 4, 1 - 2, -1)$ were used, where L performs local averaging, E acts as an edge detector and S acts as a spot detector. The following TEM features were extracted: 1) LL - texture energy (TE) from LL kernel, 2) EE - TE from EE kernel, 3) SS - TE from SS kernel, 4) LE - average TE from LE and EL kernels, 5) ES - average TE from ES and SE kernels, and 6) LS - average TE from LS and SL kernels.

7. *Fractal Dimension Texture Analysis (FDTA)* [30]: Hurst coefficient, $H^{(k)}$, for resolutions $k = 1, 2, 3, 4$.

8. *Fourier Power Spectrum (FPS)* [30]: 1) Radial sum, and 2) Angular sum.

9. *Shape Parameters*: 1) X-coordinate maximum length, 2) Y-coordinate maximum length, 3) area, 4) perimeter, 5) perimeter2/area, 6) eccentricity, 7) equivalence diameter, 8) major axis length, 9) minor axis length, 10) centroid, 11) convex area, and 12) orientation.

These texture features are usually computed on a region of interest, for example the region prescribed by the plaque contour that is automatically or manually drawn (see Sections 2.1.2 and 2.2.2).

3.1.1   DISTANCE MEASURES

Despeckle filtering and texture analysis were carried out on 440 ultrasound images of the carotid. In order to identify the most discriminant features separating the two classes under investigation i.e., asymptomatic and symptomatic ultrasound images (identifying features that have the highest discriminatory power), before and after despeckle filtering the distance between asymptomatic and symptomatic images was calculated for the set of all ultrasound images, before and after despeckle filtering for each feature as follows [33]:

$$dis_{zc} = |m_{za} - m_{zs}| \, / \sqrt{\sigma_{za}^2 + \sigma_{zs}^2} \qquad (3.1)$$

where z is the feature index, c if o indicates the original image set and if f indicates the despeckled image set, m_{za} and m_{zs} are the mean values and σ_{za} and σ_{zs} are the standard deviations of the asymptomatic and symptomatic classes respectively. The most discriminant features are the ones with the highest distance values [33]. If the distance after despeckle filtering is increased i.e.,

$$dis_{zf} > dis_{zo} \qquad (3.2)$$

then it can be derived that the classes may be better separated.

For each feature, a percentage distance was computed as:

$$feat\_dis_z = \left(dis_{zf} - dis_{zo}\right) 100. \qquad (3.3)$$

For each feature set, a score distance was computed as:

$$Score\_Dis = (1/N) \sum_{z=1}^{N} \left(dis_{zf} - dis_{zo} \right) 100 \tag{3.4}$$

where N is the number of features in the feature set. It should be noted that for all features a larger feature distance shows improvement.

Table 3.1 tabulates the results of *feat_dis_z* (3.3), and *Score_Dis* (3.4), for SF, SGLDM range of values and NGTDM feature sets for the 16 despeckle filters. The results of these feature sets are presented only, since were the ones with the best performance. The filters are categorized in linear filtering, non-linear filtering, diffusion filtering and wavelet filtering, as introduced in the companion Volume I of this book [1]. Also the number of iterations (Nr. of It.) for each filter is given, which was selected based on the speckle index (C) and on the visual evaluation of the two experts. When C was minimally changing then the filtering process was stopped. The bolded values represent the values that showed an improvement after despeckle filtering compared to the original. The last row in each sub-table shows the *Score_Dis* for all features, where the highest value indicates the best filter in the sub-table.

Additionally, a total score distance *Score_Dis_T* was computed for all feature sets shown in the last row of Table 3.1. Some of the despeckle filters, shown in Table 3.1, are changing a number of texture features, by increasing the distance between the two classes, (positive values in Table 3.1), and therefore making the identification and separation between asymptomatic and symptomatic plaques more feasible. A positive feature distance shows improvement after despeckle filtering, whereas a negative shows deterioration.

In the first part of Table 3.1 the results of the SF features are presented, where the best *Score_Dis* is given for the filter *DsFhomo* followed by the *DsFlsminsc, DsFlsmv, DsFhomog, DsFnldif, DsFwaveltc, DsFmedian,* and *DsFwiener,* with the worst *Score_Dis* given by *DsFgf4d.* All filters reduced the speckle index, C. Almost all filters reduced significantly the variance, σ^2, and the kurtosis, σ^3, of the histogram, as it may be seen from the bolded values in the first part of Table 3.1.

In the second part of the Table 3.1 the results of the SGLDM-Range of values features set are tabulated. The filters with the highest *Score_Dis* in the SGLDM range of values features set, are *DsFhomo, DsFlsminsc, DsFhmedian, DsFmedian, DsFnlocal, DsFKuwahara, DsFad,* and *DsFhomog,* whereas all other the filters (*DsFlsmv, DsFwiener, DsFwaveltc, DsFgf4d, DsFlsrad,* and *DsFnldif*) are presenting a negative *Score_Dis.* Texture features, which improved in most of the filters, are the contrast, correlation, sum of squares variance, sum average, and sum variance.

In the third part of Table 3.1, for the NGTDM feature set almost all filters showed an improvement in *Score_Dis.* Best filters in the NGTDM feature set were, the *DsFhomo, DsFlsminsc, DsFhomog,* and *DsFlsmv.* Texture features improved at most were the completion, coarseness and contrast. The completion of the image was increased by all filters.

Table 3.1: Feature Distance (see (3.3)) and *Score_dis* (see (3.4)) for SF, SGLDM range of values, and NGTDM texture feature sets between asymptomatic and symptomatic carotid plaque ultrasound images. Bolded values show improvement after despeckle filtering. (*Continues.*)

Feature	Linear Filtering			Non-Linear Filtering								Diffusion				Wavelet
	DsF lsmv	DsF wiener	DsF lsminsc	DsF ls	DsF homog	DsF gf4d	DsF median	DsF homo	DsF lmedia	DsF Kuwahara	DsF nlocal	DsF ad	DsF srad	DsF nldif	DsF ncdif	DsF waveltc
No. of iterations	2	4		3	5	4	4	4	4	2		30	50	20	20	10
Window size	5x5	5x5	5x5	5x5	5x5	5x5	5x5	5x5	5x5	5x5	5x5	-	-	-	-	5x5
SF–Statistical Features																
μ	**14**	**19**	**22**	**20**	**11**	**3**	**4**	**164**	**9**	**12**	**13**	**18**	**5**	**5**	**4**	**15**
Median	-5	-26	-17	-21	-5	-15	-5	**110**	**2**	-22	-19	-29	-31	-6	-3	-15
σ^2	**18**	**18**	**38**	**34**	**13**	-2	**7**	**140**	**9**	**11**	**17**	**9**	**10**	**7**	**6**	**18**
σ^3	**12**	**5**	**16**	**15**	**7**	-0.1	**9**	**149**	**11**	**8**	**6**	**17**	**1**	**7**	**2**	**8**
σ^4	-12	-7	-14	-11	-4	-3	-6	**117**	-13	-8	-6	-21	**2**	**6**	-3	-9
C	0.4	0.3	0.3	0.3	0.3	0.4	0.4	0.08	0.4	0.3	0.4	0.3	0.3	0.4	0.3	0.3
Score_dis	27	9	45	37	22	-17	9	680	21	1	11	-6	-13	19	6	17
SGLDM Range of Values–Spatial Gray Level Dependence Matrix																
ASM	-21	-29	-0.5	-6	-4	-8	**2**	-47	-4	-8	-6	-25	-11	-17	-19	-20
Contrast	**47**	**14**	**107**	**33**	**32**	-3	**64**	**165**	**55**	**55**	**59**	**104**	**29**	**13**	**47**	**22**
Correlation	**12**	**15**	**59**	**23**	-5	**2**	**24**	**10**	**22**	**19**	**17**	**54**	**17**	-4	**11**	-4
SOSV	**9**	**18**	**40**	**11**	**16**	-2	**10**	**101**	**15**	**7**	**11**	**9**	**3**	**8**	**9**	**20**
IDM	-50	-48	-11	-34	-29	-8	**2**	**94**	**4**	-12	-14	-54	-22	-34	-41	-43
SAV	**17**	**23**	**24**	**21**	**15**	**3**	**7**	**169**	**9**	**27**	**22**	**22**	**14**	**6**	**14**	**18**
$\sum Var$	**19**	**18**	**38**	**15**	**15**	-2	**9**	**90**	**15**	**19**	**21**	**9**	**4**	**8**	**7**	**20**
$\sum Entr$	-34	-49	-14	-19	-19	-4	**3**	-11	**12**	-21	-17	-47	-59	-30	-22	-36
Score_dis	-1	-38	243	44	21	-22	121	571	128	86	93	72	-35	-50	6	23

Table 3.1: *(Continued.)* Feature Distance (see (3.3)) and *Score_dis* (see (3.4)) for SF, SGLDM range of values, and NGTDM texture feature sets between asymptomatic and symptomatic carotid plaque ultrasound images. Bolded values show improvement after despeckle filtering.

Feature	Linear Filtering						Non-Linear Filtering						Diffusion			Wavelet
	DsF lsmv	DsF wiener	DsF lsminsc	DsF ls	DsF homog	DsF gf4d	DsF homo	DsF median	DsF hmedia	DsF Kuwahara	DsF nlocal	DsF ad	DsF srad	DsF nldif	DsF ncdif	DsF waveltc
No. of iterations	2	4	4	3	5	4	4	4	4	2	2	30	50	20	20	10
Window size	5x5	5x5	5x5	5x5	5x5	5x5	5x5	5x5	5x5	5x5	5x5	-	-	-	-	5x5
NGTDM—Neighbourhood Gray Tone Difference Matrix																
Coarseness	**30**	**4**	**87**	**6**	-16	-7	**72**	**9**	**11**	**5**	**4**	-36	5	-37	-11	-33
Contrast	7	-9	-0.3	-0.1	**0.4**	-4	**105**	**8**	**12**	-3	-4	**5**	4	-27	-19	-15
Busyness	**17**	-30	**26**	-10	**1**	-4	**48**	**8**	**7**	**2**	**6**	-14	9	-39	5	**8**
Completion	**64**	**21**	**151**	**45**	**80**	**2**	**150**	**53**	**55**	**26**	**21**	**63**	21	**18**	14	**27**
Score_dis	118	-14	264	41	66	-13	375	78	85	30	27	18	39	-85	-11	-13
Score_dis – T	144	-43	551	122	108	-52	1626	208	267	117	131	84	8	-116	1	-19

ASM: Angular 2nd moment, SOSV: Sum of squares variance, IDM: Inverse difference moment, SAV: Sum average, $\sum_{Var}$: Sum Variance. Source [7], © IEEE 2005.

Finally, in the last row of Table 3.1, the total score distance, *Score_Dis_T*, for all feature sets is shown, where best values were obtained by the filters *DsFhomo, DsFlsminsc, DsFhmedian, DsFmedian, DsFlsmv, DsFls,* and *DsFKuwahara.*

3.1.2 UNIVARIATE STATISTICAL ANALYSIS

Since texture features and image/video quality metrics are not normally distributed, the Wilcoxon rank sum test for paired samples was used. This is a non-parametric alternative for the paired samples t-test, when the distribution of the samples is not normal. The Wilcoxon test for paired samples ranks the absolute values of the differences between the paired observations in sample 1 and sample 2 and calculates a statistic on the number of negative and positive differences. If the resulting p-value is small ($p < 0.05$) then it can be accepted that the median of the differences between the paired observations is statistically significant different from 0. The Wilcoxon matched-pairs signed rank sum test was used in order to detect if for each texture feature, a significant (S) difference or not (NS), exists between the original and the despeckled images and videos at $p < 0.05$. The test was applied on all the original and despeckled images and videos of the carotid artery.

Table 3.2 shows the results of the rank sum test, which was performed on the SGLDM range of values features set of Table 3.1, for the 16 despeckle filters. The test was performed to check if significant differences exist between the features computed on the 440 original and the 440 despeckled CCA ultrasound images. Filters that resulted with the most significant number of features after despeckle filtering as shown with the score row of Table 3.2 were the following: *DsFlsmv* (7), *DsFgf4d* (6), *DsFlsminsc* (5) and *DsFnldif* (4). The rest of the filters gave a lower number of significantly different features. Features that showed a significant difference after filtering were the ASM (10), inverse difference moment (8), sum entropy (8), correlation (4), sum of squares variance (4), contrast (3), sum variance, $\sum Var(3)$ and SAV (1). These features were mostly affected after despeckle filtering and they were significantly different.

3.1.3 KNN CLASSIFIER

The statistical pattern recognition k-nearest-neighbour (kNN) classifier using the Euclidean distance with $k = 7$, was used to classify a plaque as asymptomatic or symptomatic [33]. The kNN classifier was chosen because it is simple to implement and computationally very efficient. This is highly desired due to the many feature sets and filters tested [30]. In the kNN algorithm in order to classify a new pattern, its k-nearest-neighbours from the training set, are identified. The new pattern is classified to the most frequent class among its neighbours based on a similarity measure that is usually the Euclidean distance. In this work the kNN carotid plaque classification system was implemented for values of $k = 1, 3, 5, 7$ and 9 using for input the eight texture feature sets and morphology features described above.

The leave-one-out method was used for evaluating the performance of the classifier, where each case is evaluated in relation to the rest of the cases. This procedure is characterized by no bias

Table 3.2: Wilcoxon rank-sum test for the SGLDM range of values texture features applied on the 440 ultrasound images of carotid plaque before and after despeckle filtering. The test shows with S significant difference after filtering at $p < 0.05$ and NS no significant difference after filtering at $p >= 0.05$.

Feature	Linear Filtering			Non-Linear Filtering								Diffusion				Wavelet	Score
	DsF lsmv	DsF wiener	DsF lsminsc	DsF ls	DsF homog	DsF gf4d	DsF homo	DsF median	DsF hmedian	DsF Kuwahara	DsF nlocal	DsF ad	DsF srad	DsF nldif	DsF ncdif	DsF waveltc	
ASM	S	NS	S	NS	S	S	NS	NS	NS	S	NS	S	S	S	S	S	10
Contrast	S	NS	NS	NS	NS	S	NS	NS	NS	NS	NS	NS	NS	S	NS	NS	3
Correlation	S	NS	S	NS	NS	S	NS	NS	NS	NS	NS	NS	NS	NS	S	NS	4
SOSV	S	NS	NS	S	S	S	NS	NS	NS	NS	NS	NS	NS	NS	NS	NS	4
IDM	S	NS	S	NS	NS	S	S	S	S	NS	NS	NS	NS	S	NS	S	8
SAV	NS	NS	NS	NS	NS	NS	NS	NS	NS	S	NS	NS	NS	NS	NS	NS	1
$\sum$ Var	S	NS	S	NS	NS	NS	NS	NS	NS	S	NS	NS	NS	NS	NS	NS	3
$\sum$ Entr	S	NS	S	S	NS	S	NS	NS	NS	NS	S	NS	S	S	NS	S	8
Score	7	0	5	2	2	6	1	1	1	3	1	1	2	4	2	3	

ASM: Angular 2nd moment, SOSV: Sum of squares variance, IDM: Inverse difference moment, SAV: Sum average, $\sum$ Var: Sum Variance. Score: illustrates the number of S.
Source [7], © IEEE 2005.

concerning the possible training and evaluation bootstrap sets. This method calculates the error or the classifications score by using $n - 1$ samples in the training set and testing or evaluating the performance of the classifier on the remaining sample. It is known that for large n, this method is computationally expensive. However, it is approximately unbiased, at the expense of an increase in the variance of the estimator [164]. The kNN classifier was chosen because it is simple to implement and computationally very efficient. This is highly desired due to the many feature sets and filters tested [30].

Table 3.3 shows the percentage of correct classifications score for the kNN classifier with $k = 7$ for classifying a subject as asymptomatic or symptomatic. The classifier was evaluated using the leave one out method [30], on 220 asymptomatic, and 220 symptomatic images on the original and despeckled images. The percentage of correct classifications score is given for the following feature sets: Statistical Features (SF), Spatial Gray Level Dependence Matrix Mean Values (SGLDMm), Spatial Gray Level Dependence Matrix Range of Values (SGLDMr), Gray Level Difference Statistics (GLDS), Neighborhood Gray Tone Difference Matrix (NGTDM), Statistical Feature Matrix (SFM), Laws Texture Energy Measures (TEM), Fractal Dimension Texture Analysis (FDTA), and Fourier Power Spectrum (FPS). Filters that showed an improvement in classifications success score compared to that of the original image set, were in average (last row of Table 3.4) the filter *DsFhomo* (3%), *DsFgf4d* (1%) and *DsFlsminsc* (1%).

Feature sets, which benefited mostly by the despeckle filtering were (last column in Table 3.3) the SF (11), TEM (10), SFM (5), SGLDm (4), GLDS (4) and NGTDM (4) when counting the number of cases that the correct classifications score was improved. Less improvement was observed, for the feature sets FDTA, FPS and SGLDMr. For the feature set SGLDMr better results are given for the *DsFlsminsc* filter with an improvement of 2%. This is the only filter that showed an improvement for this class of features. For the feature set TEM the filter *DsFlsmv* shows the best improvement with 9%, whereas for the FPS feature set the filter *DsFlsminsc* gave the best improvement with 5%. The filter *DsFlsminsc* showed improvement in the GLDS and NGTDM feature sets, the filter *DsFlsmv* showed improvement for the feature sets SF and TEM whereas the filter *DsFhmedian* in SFM, SF and GLDS.

3.1.4 IMAGE AND VIDEO QUALITY AND VISUAL EVALUATION

Table 3.4 tabulates the image quality evaluation metrics presented in Section 2.3 (of Volume I of this book [1]) for the 220 asymptomatic and 220 symptomatic ultrasound images between the original and the despeckled images, respectively. Best values were obtained for the *DsFnldif*, *DsFlsmv* and *DsFwaveltc* with lower MSE, RMSE, Err3, and Err4 and higher SNR and PSNR. The GAE was 0.00 for all cases, and this can be attributed to the fact that the information between the original and the despeckled images remains unchanged. Best values for the universal quality index, Q, and the structural similarity index, SSIN were obtained for the filters *DsFlsmv*, *DsFnldif*, and *DsFhmedian*.

Table 3.3: Percentage of correct classifications score for the kNN classifier with $k = 7$ for the original and the filtered image sets. Bolded values indicate improvement after despeckling.

Feature Set	No. of Feat	Original	Linear Filtering			Non-Linear Filtering								Diffusion				Wavelet	Score
			DsF lsmv	DsF wiener	DsF lsminsc	DsF ls	DsF homog	DsF gf4d	DsF homo	DsF median	DsF hmedian	DsF Kuwahara	DsF nlocal	DsF ad	DsF srad	DsF nldif	DsF ncdif	DsF waveltc	
SF	5	59	**62**	**61**	**61**	57	**63**	59	**60**	**61**	59	59	**60**	**60**	58	52	**60**	**61**	11
SGLDMm	13	65	63	62	64	65	**69**	**67**	62	64	64	64	62	61	64	**66**	63	63	4
SGLDMr	13	70	66	64	**72**	68	65	70	70	69	69	**71**	69	64	69	65	69	65	2
GLDS	4	64	63	61	**66**	64	**66**	**66**	63	**66**	63	62	52	59	61	58	61	62	4
NGTDM	5	64	63	60	**68**	**65**	**65**	57	59	63	**65**	64	61	60	63	61	**65**	62	4
SFM	4	62	62	62	60	55	**65**	68	58	**64**	**65**	**63**	61	59	**63**	56	57	55	5
TEM	6	59	**68**	**60**	52	**66**	60	**65**	**61**	**61**	**66**	58	**60**	53	**60**	**60**	59	**60**	10
FDTA	4	64	63	53	**66**	53	62	**73**	63	63	53	62	61	55	62	54	61	62	3
FPS	2	59	54	59	**64**	59	59	59	52	59	59	**60**	59	52	56	48	**60**	55	3
Average		63	63	60	**64**	62	**64**	**66**	61	63	63	63	61	58	62	58	62	61	

ASM: Angular 2nd moment, SOSV: Sum of squares variance, IDM: Inverse difference moment, SAV: Sum average, $\sum_{Var}$: Sum Variance. Score: illustrates the number of S. Source [7], © IEEE 2005.

Table 3.4: Image quality evaluation metrics computed for the 220 asymptomatic and 220 symptomatic images

Feature Set	Linear Filtering						Non-Linear Filtering					Diffusion				Wavelet
	DsF lsmv	DsF wiener	DsF lsminsc	DsF ls	DsF homog	DsF gf4d	DsF homo	DsF median	DsF hmedian	DsF Kuwahara	DsF nlocal	DsF ad	DsF srad	DsF nldif	DsF ncdif	DsF waveltc
Asymptomatic Images																
Err3	7	5	17	9	14	25	38	25	22	5	9	21	21	5	6	4
Err4	11	7	26	11	24	40	49	41	23	9	24	32	29	10	11	5
GAE	0	0	0	0	0	0	0	0	0	0	0	0	0	0	0	0
SNR	25	23	17	7	21	14	5	16	9	22	7	14	12	28	16	25
PSNR	39	36	29	29	34	27	20	29	30	30	19	28	27	41	22	39
Q	0.83	0.74	0.78	0.71	0.72	0.77	0.28	0.80	0.81	0.74	0.74	0.68	0.69	0.83	0.79	0.65
SSIN	0.97	0.92	0.88	0.93	0.97	0.88	0.43	0.94	0.96	0.89	0.81	0.87	0.89	0.97	0.81	0.9
AD	0.9	0.2	0.3	0.86	0.2	-0.67	-0.99	0.44	0.51	0.4	0.6	0.3	-11	-0.1	0.3	-0.1
SC	1.3	1.1	1.0	1.2	0.9	0.75	0.04	1.8	1.2	1.1	1.1	1.2	0.9	1.1	1.1	1.2
MD	55	35	55	59	32	98	122	54	130	54	67	55	74	129	125	55
Symptomatic Images																
MSE	33	44	374	45	110	557	1452	169	131	26	19	374	134	8	11	23
RMSE	5	6	19	16	10	23	37	13	22	6	14	19	22	3	8	5
Err3	10	9	33	22	20	43	51	25	29	8	21	31	33	5	12	6
Err4	16	11	47	41	30	63	64	38	36	5	16	43	41	7	11	8
GAE	0	0	0	0	0	0	0	0	0	0	0	0	0	0	0	0
SNR	24	22	13	22	17	12	5	16	14	28	19	12	15	29	19	25
PSNR	34	33	23	17	28	21	17	26	26	24	21	23	19	39	20	36
Q	0.82	0.71	0.77	0.74	0.77	0.75	0.24	0.79	0.80	0.78	0.78	0.63	0.71	0.81	0.77	0.49
SSIN	0.97	0.86	0.85	0.82	0.84	0.85	0.28	0.81	0.88	0.82	0.84	0.81	0.77	0.97	0.81	0.87
AD	2.1	0.9	0.5	1.9	0.5	-0.61	-0.96	0.67	0.62	0.6	0.9	0.4	-17	-0.3	03	-0.3
SC	1.5	1.3	1.4	1.7	1.0	0.98	0.05	1.4	1.4	1.3	1.6	1.4	0.7	0.9	1.4	1.4
MD	88	39	59	78	31	76	139	52	121	67	53	59	67	111	110	63

MSE: Mean square error, RMSE: Randomised mean square error, Err3, Err4: Minkowski metrics, GAE: Geometric average error, SNR: Signal to noise radio, PSNR: Peak signal to noise radio, Q: Universal quality index, SSIN: Structural similarity index, AD: Average difference, SC: Structural content, Maximum difference. Source [7], © IEEE 2005.

As presented in Section 7.3 of the companion monograph [1] (in Fig. 7.4–Fig. 7.6 and Table 7.3–Table 7.5), the best visual results as assessed by the two experts were obtained for the filters *DsFlsmv, DsFlsminsc* and *DsFkuwahara* whereas the filters *DsFgf4d, DsFad, DsFncdif,* and *DsFnldif* also showed good visual results but smoothed the image, loosing subtle details and affecting the edges. Filters that showed a blurring effect were the *DsFmedian, DsFwiener, DsFhomog,* and *DsFwaveltc.* Filters *DsFwiener, DsFls, DsFhomog,* and*DsFwaveltc* showed poorer visual results. The experts evaluated an artificial carotid artery image, a phantom ultrasound image and a real ultrasound image of the CCA.

Table 3.5A shows the results of the visual evaluation of the original and despeckled images made by two experts, a cardiovascular surgeon and a neurovascular specialist. They evaluated 100 ultrasound images before and after despeckle filtering (50 asymptomatic (A) and 50 symptomatic (S)). For each case a total of 10 images were evaluated (one original and nine filtered). For each case, for each image, the experts assigned a score in the one to five scale based on subjective criteria. Therefore, the maximum score for a filter is 500, if the expert assigned the score of five for all the 100 images. For each filter, the score was divided by five to be expressed in percentage format. The last row of Table 3.5A presents the overall average percentage (%) score assigned by both experts for each filter.

For the cardiovascular surgeon, the average score, showed that the best despeckle filter is the *DsFlsmv* with a score of 62%, followed by *DsFgf4d, DsFhmedian, DsFhomog* and *original* with scores of 52%, 50%, 45% and 41%, respectively. For the neurovascular specialist, the average score showed that the best filter is the *DsFgf4d* with a score of 72%, followed by *DsFlsmv, original, DsFlsminsc,* and*DsFhmedian* with scores of 71%, 68%, 68% and 66%, respectively. The overall average % score shows that the highest score was given to the filter *DsFlsmv* (67%), followed by *DsFgf4d* (62%), *DsFhmedian* (58%), and *original* (54%). It should be emphasized that the despeckle filter *DsFlsmv* is the only filter that was graded with a higher score than the original by both experts for the asymptomatic and symptomatic image sets.

We may observe a difference in the scorings between the two vascular specialists and this is because, the cardiovascular surgeon is primarily interested in the plaque composition and texture evaluation whereas the neurovascular specialist is interested to evaluate the degree of stenosis and the lumen diameter in order to identify the plaque contour. Filters *DsFlsmv* and *DsFgf4d* were identified as the best despeckle filters, by both specialists as they improved visual perception with overall average scores of 67% and 62%, respectively. The filters *DsFwaveltc* and *DsFhomo* were scored by both specialists with the lowest overall average scores of 28% and 29%, respectively.

Table 3.5B shows the results of the visual perception evaluation made by the same experts, one year after the first visual evaluation. The visual perception evaluation was repeated in order to assess the intra-observer variability between the same expert and was performed under the same conditions as the first visual evaluation.

For the cardiovascular surgeon, the average score, showed that the best despeckle filter is again the *DsFlsmv* with a score of 61%, followed by *DsFhmedian, DsFgf4d, DsFls, DsFhomog* and

Table 3.5: A. Percentage scoring of visual evaluation of the original and despeckled images (50 asymptomatic (A) and 50 symptomatic (S)) by the experts [11]

Experts	A/S	DsForiginal	Linear Filtering		Non-Linear Filtering				Diffusion	Wavelet
			DsFlsmv	DsFlsminsc	DsFlhmedian	DsFhomog	DsFgf4d	DsFhomo	DsFnldif	DsFwaveltc
Cardiovascular	A	33	75	33	43	47	61	19	43	32
surgeon	S	48	49	18	57	43	42	20	33	22
Average %		41	62	26	50	45	52	19	38	27
Neurovascular	A	70	76	73	74	63	79	23	52	29
specialist	S	66	67	63	58	45	65	55	41	28
Average %		68	71	68	66	54	72	39	47	28
Overall Average %		54	67	47	58	50	62	29	43	28

Source © [11]

Table 3.5: B. Percentage scoring of visual evaluation of the original and despeckled images (50 asymptomatic (A) and 50 symptomatic (S)) by the experts one year after the first visual evaluation [11]

Experts	A/S	original	Linear Filtering			Non-Linear Filtering				Diffusion	Wavelet
			DsFlsmv	DsFlsminsc	DsFlmedian	DsFls	DsFhomog	DsFgf4d	DsFhomo	DsFnldif	DsFwaveltc
Cardiovascular	A	28	57	43	62	49	41	53	16	39	31
surgeon	S	44	65	24	57	49	39	51	23	37	21
Average %		36	61	34	60	49	40	52	20	38	26
Neurovascular	A	62	65	64	69	67	51	65	19	49	24
expert	S	64	62	71	53	51	49	69	49	44	26
Average %		63	64	68	61	59	50	67	34	47	25
Overall Average %		50	63	51	61	54	45	60	27	43	26

Source © [11]

original with scores of 60%, 52%, 49%, 40% and 36%, respectively. For the neurovascular expert, the average score showed that the best filter is the *DsFlsminsc* with a score of 68%, followed by *DsFgf4d, DsFlsmv, original,* and *DsFhmedian* with scores of 67%, 64%, 63% and 61%, respectively. The overall average % score shows that the highest score was given to the filter *DsFlsmv* (63%), followed by *DsFhmedian* (61%), *DsFgf4d* (60%), *DsFls* (54%), and *original* (50%). The intra-observer variability results in Table 3.5B shows a consistency in almost all results, with only very small differences between filters. The despeckle filter *DsFlsmv* is again, the only filter that was graded with a higher score than the original by both vascular experts for the asymptomatic and symptomatic images.

Both experts were in agreement that the best despeckle filters for visual perception, are the *DsFlsmv, DsFlsminsc, DsFgf4d,* and *DsFhmedian,* whereas the worst filters were the *DsFwaveltc* followed by the *DsFhomo* and *DsFnldif* (see also Table 3.5A, and Table 3.5B). Furthermore, both experts agreed that almost all despeckle filters reduced the noise substantially and images may be better visualized after despeckle filtering. By examining the statistical results of Tables 3.1–Table 3.4 and the visual evaluation of Table 3.5 we can conclude that the best filters are *DsFlsmv* and *DsFgf4d,* which may be used for both plaque composition enhancement and plaque texture analysis, whereas the filters *DsFlsmv, DsFgf4d* and *DsFlsminsc* are more appropriate to identify the degree of stenosis and therefore may be used when the primary interest is to outline the plaque borders.

3.2   DISCUSSION OF IMAGE DESPECKLE FILTERING BASED ON TEXTURE ANALYSIS

The results on texture analysis, presented in Table 3.2–Table 3.4, showed that the filters *DsFlsmv, DsFgf4d,* and *DsFlsminsc,* improved the class separation between the asymptomatic and the symptomatic classes by increasing the distance between them. These filters, *DsFlsmv, DsFgf4d,* and *DsFlsminsc,* gave the highest number of significantly different features (Table 3.2), with 7, 6, and 5, respectively, and gave only a marginal improvement in the percentage of correct classification success rate (Table 3.3). The high number of significantly different features for these filters, showed that the two classes (asymptomatic, symptomatic) may be better separated after despeckle filtering with the filters *DsFlsmv, DsFgf4d,* and *DsFlsminsc.* Table 3.1 showed that almost all despeckle filters increased the distance between the asymptomatic and the symptomatic images thus making the identification of a class more easily to identify. Table 3.1 also showed that most of the filters reduced the asymmetry, σ^3 and the skewness, σ^4 of the histogram. Table 3.2 showed, that despeckle filtering influenced more some statistical features, such as the inverse difference moment, IDM, the angular second moment, ASM, and the sum entropy, $\sum Entr$, while other statistical features were less influenced by despeckle filtering. As a result, these features, which were more influenced, may be used in future research to evaluate despeckle filtering. The *Score_Dis_T* in the last row of Table 3.2, showed that best feature distance was given by the filters *DsFhomo, DsFlsminsc, DsFmedian,* and *DsFlsmv.* Table 3.3 showed that not all feature sets equally benefited from despeckle filtering. Specifically, the SF and TEM feature sets benefited from almost all despeckle filters (7), whereas the feature sets SGLDMm, GLDS, and NGTDM, benefited from four despeckle filters, FDTA three and SFM two. The features sets SGLDMr, and FPS, benefited from only one despeckle filter.

There were some results given in the recent literature based on texture analysis of ultrasound images for, the classification of atherosclerotic carotid plaque [7, 26, 33, 154], liver ultrasound images [30], electron microscopic muscle images [155], detection of breast masses [156], cloud images [100] SAR images [41, 157], and some results given on artificial images from the pioneer researchers in texture analysis [152, 153]. There is no other study reported in the literature, where texture analysis (Table 3.1–Table 3.3) was used to the extent, that is used in our study, to evaluate despeckle filtering in ultrasound imaging. In studies [100, 144], some of the texture measures used in our study (Table 3.3), were also used on a total of 230 ultrasound images of the carotid plaque (115 asymptomatic, 115 symptomatic), in order to characterise carotid plaques as safe or unsafe and identify patients at risk of stroke. Specifically in [30] and [100] all nine different features used in our study (see Table 3.3) were also used to classify a plaque as asymptomatic or symptomatic, where comparable values as in our study were obtained for all feature sets. Examples of the use of texture analysis were also provided in [156], for classifying malignant and benign tumours of breast, in [100] for classifying clouds and predicting weather, and finally in [153] to automatically classify terrain texture.

Despeckle filtering was investigated by other researchers and also in our study, on an artificial carotid image (Fig. 7.1, vol. I), [1]–[3], [7, 24], [11]–[14], on line profiles (Fig. 7.1, vol. I) of different ultrasound images, [1, 7, 11, 26, 46, 51, 56], on phantom ultrasound images (see also Fig. 7.3, vol. I) [28, 51, 54], SAR images [1, 2, 41, 157, 158] real longitudinal ultrasound images of the carotid artery (Fig. 7.4, vol. I) [1, 2, 7, 11, 24, 51, 159] and cardiac ultrasound images (see Fig. 7.5, vol. I). There are only few studies [2, 7, 11, 24] where despeckle filtering was investigated on real, and artificial longitudinal ultrasound image of the carotid artery. Four different despeckle filters were applied in [24], namely the *DsFlsmv*, [19], Frost [18], *DsFad* [50], and a *DsFsrad* filter [24]. The despeckle window used for the *DsFlsmv*, and Frost filters was 7×7 pixels. To evaluate the performance of these filters, the mean and the standard deviation were used, which were calculated in different regions of the carotid artery image, namely in lumen, tissue, and at the vascular wall. The mean gray level values of the original image for the lumen, tissue and wall regions were 1.03, 5.31, and 22.8, whereas the variance were 0.56, 2.69, and 10.61. The mean after despeckle filtering with the *DsFsrad* gave brighter values for the lumen and tissue. Specifically the mean for the lumen, tissue, and wall for the *DsFsrad* was (1.19, 6.17, 18.9), *DsFlsmv* (1.11, 5.72, 21.75), Frost (1.12, 5.74, 21.83) and *DsFad* (0.90, 4.64, 14.64). The standard deviation for the *DsFsrad* gave lower values (0.15, 0.7, 2.86) when compared with Lee (0.33, 1.42, 5.37), Frost (0.32, 1.40, 5.30), and *DsFad* (0.20, 1.09, 3.52). It was thus shown that the *DsFsrad* filter preserves the mean and reduces the variance. The number of images investigated in [24], was very small, visual perception evaluation by experts was not carried out, as well as only two statistical measures were used to quantitatively evaluate despeckle filtering, namely the mean, and the variance before and after despeckle filtering as explained above. We believe that the mean and the variance used in [24] are not indicative and may not give a complete and accurate evaluation result as in [7]. Furthermore, despeckle filtering was investigated by other researchers on ultrasound images of, heart [51] (see also Fig. 7.5, vol. I) [1]), pig heart [56], pig muscle [158], kidney [54], liver [160], echocardiograms [55], CT lung scans [48], MRI images of brain [161], brain X-ray images [46] SAR images [41], and real world images [162].

Line plots, as used in our study (see Fig. 7.2, vol. I [1] and [11]), were also used in few other studies to quantify despeckle filtering performance. Specifically in [46], a line profile through the original and the despeckled ultrasound image of kidney was plotted, using adaptive Gaussian filtering. In [177] line profiles were plotted on four simulated and 15 ultrasound cardiac images of the left ventricle, in order to evaluate the *DsFmedian* filter. In another study [51], line profiles through one phantom, one heart, one kidney, and one liver ultrasound image, were plotted where an adaptive shrinkage weighted median [160, 162], *DsFwaveltc* (wavelet shrinkage) [57], and wavelet shrinkage coherence enhancing [55] models were used and compared with a nonlinear coherent diffusion model [57]. Finally in [56], line plots were used in one artificial computer simulated image, and one ultrasound image of pig heart, where an adaptive shrinkage weighted median filter [160, 162], a multiscale nonlinear thresholding without adaptive filter pre-processing [56], a wavelet shrinkage filtering method [57], and a proposed adaptive nonlinear thresholding with adaptive pre-processing method [56], were evaluated. In all of the above studies, visual perception evaluation by experts, statistical and texture analysis, on multiple images, as performed in our study, was not performed.

Phantom images were used in this book (see Fig. 7.3, vol. I, [1]) and by other researchers to evaluate despeckle filtering in carotid ultrasound imaging. Specifically in [51], a synthetic carotid ultrasound image of the CCA was used to evaluate the *DsFsard* filtering (speckle reducing anisotropic filtering) which was compared with the *DsFlsmv* (Lee filter) [19], and the *DsFad* filter (conventional anisotropic diffusion) [49]. The edges of the phantom image used in [51], were studied and it was shown that the *DsFsrad* does not blurred edges as with the other two despeckle filtering techniques evaluated (*DsFlsmv* and *DsFad*).

While there are a number of despeckle filtering techniques and commercial software packages proposed in the literature for despeckling of ultrasound images which are presented in [1, 2] we found no other studies in the literature for despeckle filtering in ultrasound videos of the CCA with the exception of [3]. More specifically, we proposed in [3], a freeware despeckle filtering toolbox which was based on four despeckle filtering methods (*DsFlsmv, DsFhmedian, DsFkuwahara,* and *DsFsrad*) for video despeckling. A number of studies have investigated additive noise filtering in natural video sequences [68]–[71], [74] and [75].

3.3 DISCUSSION OF IMAGE DESPECKLE FILTERING BASED ON VISUAL QUALITY EVALUATION

The visual perception evaluation performed in Table 3.5A and Table 3.5B, showed that the filters *DsFlsmv*, *DsFgf4d*, and *DsFlsminsc* improved the visual assessment by experts. The intra-observer variability test (Table 3.5A), which was repeated one year after the first visual evaluation (Table 3.5B), showed that the differences between the visual evaluations made by the two experts were very small, and the results of the two tables were in agreement.

It was shown that the highest scores were obtained, for the filter *DsFlsmv* for both tables. The differences, which are observed in the ratings between the two experts, were due to the fact that each expert was interested for a different tissue area in the ultrasound image of the carotid artery. Specifically the cardiovascular surgeon was primarily interested in the plaque composition and texture, whereas the neurovascular expert was interested in the degree of stenosis and the lumen diameter. The filter *DsFlsminsc* was rated from the neurovascular expert with the highest score in Table 3.5B. The expert found that this filter was very helpful when inspecting the degree of stenosis and the lumen diameter.

Furthermore, the two experts, evaluated the images before and after despeckle filtering, and gave some additional comments, which we think it is important to briefly discuss. It was shown that the primary interest of the experts were the borders between IMT, plaque, artery wall, and blood, in order to be able to exactly make a separation between them. Other important points taken into consideration from both experts during this examination were the texture of plaque, as the texture may give indication about the risk of stroke [33]. They have both commented the fact that the *DsFlsmv* filter was good for visualising the borders between blood, plaque and wall but not between wall and surrounding tissue, the *DsFlsminsc* helped specifically for the plaque visualisation as plaque borders were better after filtering, and that the *DsFgf4d* sharpened the edges, thus it may be used for plaque visualization and to separate the borders between blood and plaque.

To the best of our knowledge, no other studies were carried out, (with the exception of [7] and [8]) on the visual evaluation of ultrasound images by using despeckle filtering and image normalisation with two [46] experts. More specifically in [7], 56 different textures features and 10 different image quality evaluation metrics were used to compare the effect of despeckle filtering in 440 ultrasound images of the carotid artery, where two different experts optically evaluated the images. It was found that a linear order statistics filter, based on first order statistics may be successfully used for despeckling carotid artery ultrasound images. Furthermore, in [8], two different ultrasound imaging scanners, namely the ATL HDI-3000 and the ATL HDI-5000 were compared based on texture features and image quality metrics extracted form 80 ultrasound images of the carotid bifurcation, before and after despeckle filtering. It was shown that normalisation and despeckle filtering favours image quality. In a significant number of despeckle filtering studies, [19, 24, 41, 47, 51, 55, 56], [158]–[163] visual evaluation was carried out by non-experts. There are very few results reported in the literature, where visual perception evaluation was carried out in ultrasound images. Specifically, despeckle filtering was evaluated visually by two experts in [55], where they manually delineated 60 echocardiographic images before and after despeckle filtering. Quantitative measurements were calculated in terms of the mean of absolute border difference and the mean of border area differences. The visual evaluation in [55], showed that the borders, which were manually defined by the experts were improved after despeckle filtering. In [164], the performance assessment of multi-temporal SAR image despeckling was evaluated from ten photo interpreters. The evaluation was made between the original and the three filtered results. The photo interpreters evaluated the accuracy of manual detection of geographical features, such as lines points and surfaces, by presenting the images in random order. The ten photo interpreters concluded that despeckle filtering improves the identification of the above criteria and that specific filters may be used to enhance points, lines or surfaces as required. In another study image quality was evaluated for compressed still images [165], where the images were presented to an unknown number of observers in random order. The observers were not experts, but they were untrained persons over 18 drawn from the university population.

The image quality evaluation results presented in Table 3.4 showed that the best values were obtained by the despeckle filters *DsFnldif, DsFlsmv*, and *DsFwaveltc*. It was shown from Table 3.4, that the effect of despeckle filtering was more obvious on the asymptomatic images, where generally better image quality evaluation results were obtained. Moreover, it is obvious that all quality evaluation metrics presented here were equally important for image quality evaluation. It is furthermore important to note that a higher PSNR (or equivalently, a lower RMSE) does not necessarily imply a higher subjective image quality, although they do provide some measure of relative quality. While some quality metrics for different images have been studied and proposed in the literature, such as for MRI [166], natural and artificial images [4], to the best of our knowledge, no other comparative study exists except [7], which have investigated the application of the above metrics together with visual perception evaluation, on ultrasound images of the carotid artery. In previous studies [6, 17, 19, 41, 42] researchers evaluated image quality on real world images using either only the visual perception by experts or some of the evaluation metrics presented in Table 3.4. In all these studies, the comparison of the proposed method was made with another one, based on image quality evaluation metrics, such as the MSE [1–3, 7, 46, 51, 54, 56, 162], PSNR [162], SNR [7, 46], C [7, 54], the mean, and the variance [7, 24, 41, 158, 160] and line plots [7, 19, 51, 56, 158] between the original and despeckled images. The usefulness of these measures was not investigated for the despeckling of ultrasound images. Furthermore, normalization and despeckling was not taken into consideration as in our study. In a recent study [8], we have investigated the image quality on ultrasound images of the carotid artery, where it was shown that despeckle filtering, increases the quality of these images and also increases the accuracy of the IMT [98] and plaque [9] segmentation.

Image quality metrics were also investigated for the evaluation of ultrasound spatial compound scanning [167], to compare the quality of JPEG images before and after compression using the PSNR, and SSIN [4], where values for the PSNR, and SSIN of 8.45, and 0.96, were measured respectively, while in our study, we have achieved values of 39, and 0.97, with the *DsFlsmv* filter (see Table 3.5). In [31], real world images were evaluated based on their compression ratios, by using the MSE, and Q, where values of 30, and 0.92 were reported, respectively. Furthermore, real world images were also evaluated in [31], based on the MSE and Q before and after, histogram equalization (1144.2, 0.74), median filtering (14.47, 0.78), wavelet compression (16.03, 0.68), and spatial displacement (141.2, 0.5).

In another study [162], where various median filtering techniques were investigated on real world images, the image quality measures, MSE, and PSNR, were used to compare, between the original and the filtered images. In [168] number of quality metrics were reviewed to evaluate JPEG compression on still real world images, such as the MSE, SNR, PSNR, M_3, and M_4. In [51], where despeckle filtering was investigated on artificial and ultrasound images of heart, kidney and abdomen, the MSE values reported after despeckle filtering were 289, 271, 132, and 121, for four different despeckle filtering methods, namely the adaptive weighted median filtering [162], wavelet shrinkage enhanced [55], wavelet shrinkage [50], and non-linear coherence diffusion method [57]. Most of the researchers used the image quality measures such as the MSE [7, 51, 54, 56, 162], SNR [7, 22, 26, 53, 59], and PSNR [162], in order to compare the original with the despeckled images.

In A. Achim's et al. [54] research, values reported for the MSE were 133, 43, 49, 26, 22 for the original, and four despeckled SAR images, respectively. In Achim's research four different despeckling methods were used, namely the Lee [6], gamma MAP filter [158], soft thresholding, and the WIN-SAR filter [54], which used a 7×7 pixel filtering window, and were applied on real world and SAR images.

In another study [54], MSE values reported were 26 for the original kidney ultrasound image, 13.7 after despeckling by median filtering [162], 13.8 after homomorphic *DsFwiener* filtering [41], 13.6 after soft thresholding [57], 13.5 after hard thresholding [57], and 12.74 after Bayesian denoising [54]. In our study the MSE values for the filter *DsFlsmv*, *DsFwiener*, *DsFnldif*, and *DsFwaveltc*, (Table 3.5) were 13, 19, 8, 11, for the asymptomatic, and 33, 44, 8, 23, for the symptomatic images respectively, which are better or comparable with other studies reported above.

Normalization and speckle reduction filtering are very important preprocessing steps in the assessment of atherosclerosis in ultrasound imaging. The usefulness of image quality evaluation, in 80 ultrasound images of the carotid bifurcation, based on image quality metrics and visual perception after normalization and speckle reduction filtering using two different ultrasound scanners (ATL HDI-3000, and ATL HDI-5000) was addressed in volume I of this book [1], and furthermore here in this chapter discussed (see Section 3.6). Specifically, the images were evaluated, before, and after speckle reduction, after normalization, and after normalization and speckle reduction filtering (see Fig. 3.1 and Fig. 3.2). The evaluation was based on visual evaluation by two experts (see Table 3.8), statistical and texture features (see Table 3.11–Table 3.13), image normalization, speckle reduction, as well as based on image quality evaluation metrics (see Table 3.14).

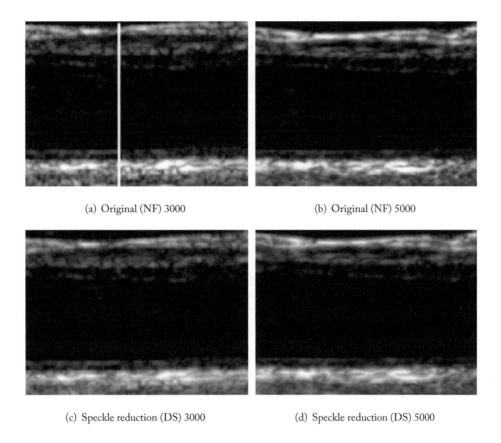

(a) Original (NF) 3000

(b) Original (NF) 5000

(c) Speckle reduction (DS) 3000

(d) Speckle reduction (DS) 5000

Figure 3.1: Ultrasound carotid artery images of the original (NF), despeckle (DS), normalized (N), and normalized despeckled (NDS), of the ATL HDI-3000, and ATL HDI-5000 shown in the left and right columns, respectively. Vertical lines given in the original image (NF) of the ATL HDI-3000 and the ATL HDI-5000 scanners, define the position of the gray-value line profiles plotted in Fig. 3.2. Source [8], © MBEC 2006. *(Continues.)*

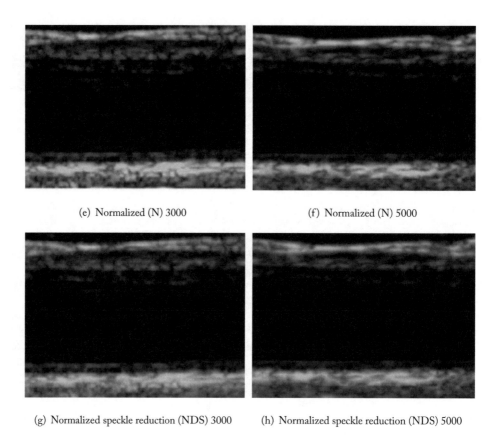

(e) Normalized (N) 3000   (f) Normalized (N) 5000

(g) Normalized speckle reduction (NDS) 3000   (h) Normalized speckle reduction (NDS) 5000

Figure 3.1: *(Continued.)* Ultrasound carotid artery images of the original (NF), despeckle (DS), normalized (N), and normalized despeckled (NDS), of the ATL HDI-3000, and ATL HDI-5000 shown in the left and right columns, respectively. Vertical lines given in the original image (NF) of the ATL HDI-3000 and the ATL HDI-5000 scanners, define the position of the gray-value line profiles plotted in Fig. 3.2. Source [8], © MBEC 2006.

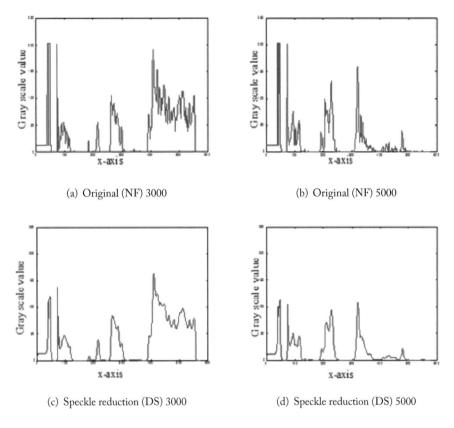

(a) Original (NF) 3000

(b) Original (NF) 5000

(c) Speckle reduction (DS) 3000

(d) Speckle reduction (DS) 5000

Figure 3.2: Gray-value line profiles of the lines illustrated in Fig. 3.1a and b, for the NF, DS, N, and NDS images, for the ATL HDI-3000, and ATL HDI-5000 scanner, shown in the left and right columns, respectively. The gray-scale value, and the column 240, is shown in the y- and x-axis. Source [8], © MBEC 2006. *(Continues.)*

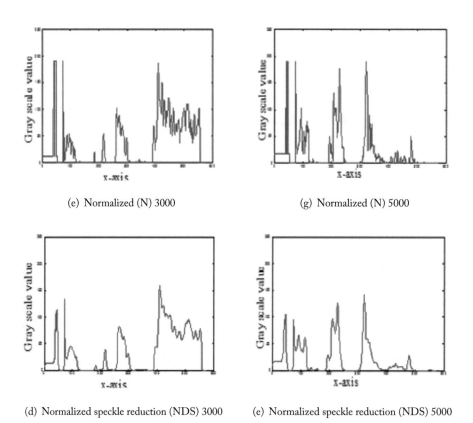

(e) Normalized (N) 3000   (g) Normalized (N) 5000

(d) Normalized speckle reduction (NDS) 3000   (e) Normalized speckle reduction (NDS) 5000

Figure 3.2: *(Continued.)* Gray-value line profiles of the lines illustrated in Fig. 3.1a and b, for the NF, DS, N, and NDS images, for the ATL HDI-3000, and ATL HDI-5000 scanner, shown in the left and right columns, respectively. The gray-scale value, and the column 240, is shown in the y- and x-axis. Source [8], © MBEC 2006.

It is noted that to the best of our knowledge, there are no other studies found in the literature for evaluating ultrasound image quality, based on speckle reduction filtering and normalization performed on carotid artery images, acquired by two different ultrasound scanners.

The main findings of this study can be summarized as follows: 1) the NDS, images were rated visually better on both scanners, 2) the NDS images showed better statistical and texture analysis results on both scanners, 3) better image quality evaluation results were obtained between the NF-N images for both scanners, followed by the NF-DS images for the ATL HDI-5000 scanner and the NF-DS on the HDI ATL-3000 scanner, 4) the ATL HDI-5000 scanner images have considerable higher entropy than the ATL HDI-3000 scanner and thus more information content. However, based on the visual evaluation by the two experts, both scanners were rated similarly.

It was shown that normalization and speckle reduction produces better images. Normalization was also proposed in other studies using blood echogenicity as a reference and applied in carotid artery images [167]. In [5, 8, 9] it was shown that normalization improves the image comparability by reducing the variability introduced by different gain settings, different operators, and different equipment. It should be noted that the order of applying these processes (normalization and speckle reduction filtering) affects the final result. Based on unpublished results, we have observed that by applying first speckle reduction filtering and then normalization produces distorted edges. The preferred method is to apply first normalization and then speckle reduction filtering for better results.

In two recent studies [9, 98] it was shown that the preprocessing of ultrasound images of the carotid artery with normalization and speckle reduction filtering improves the performance of the automated segmentation of the intima media thickness [98] and plaque [9]. More specifically, it was shown in [48] that a smaller variability in segmentation results was observed when performed on images after normalization and speckle reduction filtering, compared with the manual delineation results made by two medical experts. Furthermore, in another study [7], we have shown that speckle reduction filtering improves the percentage of correct classifications score of symptomatic and asymptomatic images of the carotid. Speckle reduction filtering was also investigated by other researchers on ultrasound images of liver and kidney [169], and on natural scenery [19], using an adaptive two-dimensional filter similar to the *DsFlsmv* speckle reduction filter used in this study. In these studies, [19, 169] speckle reduction filtering was evaluated based only on visual perception evaluation made by the researches.

Verhoefen et al. [170] applied mean and median filtering in simulated ultrasound images and in ultrasound images with blood vessels. The lesion-signal-to-noise ratio was used in order to quantify the detectability of lesions after filtering. Filtering was applied on images with fixed and adaptive size windows in order to investigate the influence of the filter window size. It was shown that the difference in performance between the filters was small but the choice of the correct window size was important. Kotropoulos et al. [171] applied adaptive speckle reduction filtering in simulated tissue mimicking phantom, and liver ultrasound B-mode images, where it was shown that the proposed maximum likelihood estimator filter was superior to the mean filter.

Although in this study, speckle has been considered as noise, there are other studies where speckle, approximated by the Rayleigh distribution, was used to support automated segmentation. Specifically, in [150], an automated luminal contour segmentation method based on a statistical approach, was introduced whereas in [172], ultrasound intravascular images were segmented using knowledge-based methods. Furthermore, in [151] a semi-automatic segmentation method for intravascular ultrasound images, based on gray-scale statistics of the image was proposed, where the lumen, IMT and the plaque were segmented in parallel by utilizing a fast marching model.

Some statistical measures, as shown in the upper part of Table 3.11, were better after normalization and some others, shown in the bottom part of Table 3.11, were better after speckle reduction. Table 3.11 also shows that the contrast was higher for the NF and N images on both scanners and was significantly difference (S) after normalization and speckle reduction filtering (see Table 3.12). All other measures presented in Table 3.11 were comparable showing that better values were obtained on the NDS images. Moreover, it was shown that the entropy that is a measure of the information content of the image [152] was higher for both scanners in the cases of the NDS and DS images. Significantly different entropy values were obtained mostly after normalization and speckle reduction filtering (see Table 3.12).

Low entropy images have low contrast and large areas of pixels with same or similar gray level values. An image which is perfectly flat will have a zero entropy. On the other hand, high entropy images have high contrast and thus higher entropy values [40]. The ATL HDI-5000 scanner produces therefore images with higher information content. The entropy was also used in other studies to classify the best liver ultrasound images [30], where it was shown that the experts rated images with higher entropy values better. In [33] entropy and other texture features were used to classify between symptomatic and asymptomatic carotid plaques for assessing the risk of stroke. It was also shown [34] that, asymptomatic plaques tend to be brighter, have higher entropy and more coarse, whereas symptomatic plaques tend to be darker, have lower entropy (i.e., the image intensity in neighbouring pixels is more unequal) and are less coarse. Furthermore, it is noted that texture analysis could also be performed on smaller areas of the carotid artery, such as the plaque, after segmentation [9, 98].

In previous studies [4, 32, 173–177] researchers evaluated image quality on natural scenery images using either only the visual perception by experts or some of the evaluation metrics presented in Table 3.13. In this study, MSE and RMSE values were in the range of 0.4 to 2.0, for all cases, Err3, Err4, SNR, PSNR, Q, and SSIN were better between the NF-N images for both scanners, showing that normalization increases the values of these measures. In [51], speckle reduction filtering was investigated on ultrasound images of the heart. The MSE values reported after speckle reduction for the adaptive weighted median filtering, wavelet shrinkage enhanced filter, wavelet shrinkage filter, and non-linear coherence diffusion were 289, 271, 132, and 121, respectively. Loupas et al. [76] applied an adaptive weighted median filter for speckle reduction in ultrasound images of the liver and gallbladder and used the speckle index and the MSE for comparing the filter with a conventional mean filter. It was shown that the filter improves the resolution of small structures in the ultrasound images. It was also documented in [4] that the MSE, RMSE, SNR and PSNR measures are not objective for image quality evaluation and that they do not correspond to all aspects of the visual perception nor they correctly reflect artifacts [32].

Recently the Q [173], and SSIN [4], measures for objective image quality evaluation have been proposed. The best values obtained in this study were Q=0.95 and SSIN=0.95 and were obtained for the NF-N images for both scanners. These results were followed with Q=0.73, and SSIN=0.92 in the case of NF-NDS for the HDI ATL-3000 scanner, and Q=0.72, and SSIN=0.94 in the case of NF-DS for the HDI ATL-5000 scanner. In [173], where natural scenery images were distorted by speckle noise, the values for Q reported were 0.4408, whereas the values for Q after contrast stretching were 0.9372.

The methodology presented in this study may also be applicable in future studies, to the evaluation of new ultrasound and telemedicine systems in order to compare their performance. It is also important to note that the methodology consists of a combination of subjective and objective measures that should be combined together for a proper image quality evaluation result [32].

3.4 EVALUATION OF DESPECKLE FILTERING ON CAROTID PLAQUE VIDEO BASED ON TEXTURE ANALYSIS

Despeckle filtering was evaluated on 10 videos of the CCA where texture features were extracted from the original and the despeckled videos from the whole video and the ROI (see Fig. 2.7– Fig. 2.9). Table 3.6 presents the results of selected statistical features (from the SF and SGLDM feature sets, see Section 3.1), that showed significance difference after despeckle filtering ($p <$ 0.05). The features were extracted from the original video frame and the despeckled video frames for the whole video frame and the ROI, for all 10 videos investigated. These features were the median, variance, SOV, IDM, entropy, DE and coarseness. It is shown that the filters *DsFlsmv* and *DsFhmedian* comparatively preserved the features median, variance and entropy but increased coarseness. It should be noted that these findings cannot be compared with the results presented in [7] and [8], as the texture features in these two studies were computed for the whole despeckled plaque images (and not the ROIs as defined in this examples).

Table 3.7 tabulates selected video quality metrics between the original and the despeckled videos for whole frame filtering and when the filtering was applied on an ROI. It is clearly shown that the despeckle filter *DsFlsmv* performs better in terms of quality evaluation for the metrics SSI, VSNR, IFC, NQM, and WSNR when applied on the whole frame. Moreover, all the investigated evaluation metrics gave better results when the *DsFlsmv* was applied only on the ROI, followed by the *DsFhmedian*.

Table 3.8 presents the results of the visual evaluation of the original and despeckled videos made by the two experts (Fig. 2.6–Fig. 2.8), a cardiovascular surgeon (Expert 1) and a neurovascu-

Table 3.6: Texture features (Mean±SD), that showed significant difference (using the Wilcoxon rank–sum test at $p < 0.05$) after despeckle filtering, for all 10 videos of the CCA extracted from the original and the despeckled videos from the whole video and the ROI (-/-)

Features	original	DsFlsmv	DsFhmedian	DsFkuwahara	DsFsrad
Median	43±14/23±17	43±14/28±18	43±14 26±12	42±14/ 26±17	43±14/26±14
Variance	54±7/58±8	53±6/58±9	54±6/58±8	55±8/ 59±9	54±6/62±8
SOV	8±6/7±4	12±11/9±3	8±7/14±22	11±7/ 7±5	10±9/12±13
IDM	0.27±0.07/	0.29±0.07/	0.39±0.05/ 0.48±0.09	0.41±0.07/	0.38±0.06/
	0.29±0.14	0.43±0.09		0.51±0.089	0.38±0.094
Entropy	7.8±0.5/7±0.96	7.7±0.4/6.7±0.9	7.5±0.4/ .6±0.9	7.5±0.6/6.6±1.1	7.4±0.5/7.01±0.8
DE	0.74±0.15/0.9±0.2	0.74±0.12/0.77±0.15	0.69±0.12/ 0.74±0.13	0.7±0.11/0.72±0.1	0.61±0.09/ 0.56±0.16
Coarseness	38±7/61±11	93±13/110±22	52±10/ 84±21	37±4/55±12	54±22/33±18

IQR: Inter-quartile range, SOV: Sum of squares variance, IDM: Inverse difference moment, DE: Difference entropy. Source [14], © IEEE 2012

Table 3.7: Video quality metrics (Mean±SD) for all 10 videos of the CCA extracted between the original and the despeckled videos from the whole video and the ROI (-/-)

Features	DsFlsmv	DsFlhmedian	DsFkuwahara	DsFsrad
SSI	0.98±0.01/0.98±0.05	0.97±0.001/0.96±0.06	0.77±0.025/0.84±0.03	0.96±0.025/0.88±0.08
VSNR	36±3.77/41±3.0	30±1.86/38±5.3	15±1.1/24.7±2.3	37±5.7/32±10
IFC	7.2±0.93/6.2±0.98	6.1±0.6/4.6±1.3	1.9±0.08/1.4±0.21	6.6±2.3/3.7±2.1
NQM	35.3±1.9/29±1.9	34±1.4/26.7±5.1	17.7±1.1/14.2±1.3	34.8±4.9/ 24.1±6.7
WSNR	34.8±1.8/38±0.91	33.1±1.1/35±5.4	18.8±0.9/20.6±1.3	39.1±5.2/25±8
PSNR	39.6±2.1/42.9±1.6	38.9±1.1/40.1±4.5	29.1±1.1/29.9±1.2	43.9±4.3/34.9±6.9

SSI: Structural similarity index, VSNR: Visual signal-to-noise radio, IFC: Information fidelity criterion, NQM: Noise quality measure, WSNR: Weighted signal-to-noise ratio, PSNR: Peak signal to noise ratio. Source [14], © IEEE 2012.

lar specialist (Expert 2). The evaluation was performed on both the whole despeckle video frame as well as to the ROI, where both methods gave similar visual evaluation scorings. The last two rows of Table 3.8 present the overall average percentage (%) score assigned by both experts for each filter and the filter ranking. It is shown in Table 3.8 that marginally the best video despeckle filter is the *DsFlsmv* with a score of 74%, followed by the filter *DsFhmedian* and *DsFkuwahara* with scores of 73% and 71%, respectively. It is interesting to note that the three filters, *DsFlsmv*, *DsFhmedian* and *DsFkuwahara*, were scored with high evaluation markings by both experts. The filter *DsFsrad* gave poorer performance with an average score of 56%.

Table 3.8: Percentage scoring of the original and despeckle videos by the experts

Experts	Original	*DsFlsmv*	*DsFhmedian*	*DsFkuwahara*	*DsFsrad*
Expert 1	33	75	71	65	61
Expert 2	40	72	75	77	51
Average %	37	74	73	71	56
Ranking	-	1	2	3	4

Source [14], © IEEE 2012

3.5   DISCUSSION OF VIDEO DESPECKLE FILTERING BASED ON TEXTURE ANALYSIS AND VISUAL QUALITY EVALUATION

Most of the papers published in the literature for video filtering are limited to the reduction of additive noise, mainly by frame averaging. More specifically, in [179] the Wiener filtering method was applied to 3D image sequences for filtering additive noise, but results have not been thoroughly discussed and compared with other methods. The method was superior when compared to the purely temporal operations implemented earlier [180]. The pyramid thresholding method was used in [180], and wavelet based additive denoising was used in [181] for additive noise reduction in image sequences. In another study [182], the image quality and evaluation metrics, were used for evaluating the additive noise filtering and the transmission of image sequences through telemedicine channels. An improvement of almost all the quality metrics extracted from the original and processed images was demonstrated. An additive noise reduction algorithm, for image sequences, using variance characteristics of the noise was presented in [183]. Estimated noise power and sum of absolute difference employed in motion estimation were used to determine the temporal filter coefficients. A noise measurement scheme using the correlation between the noisy input and the noise-free image was applied for accurate estimation of the noise power. The experimental results showed that the proposed noise reduction method efficiently removes noise. An efficient method for movie denoising that does not require any motion estimation was presented in [184]. The method was based on the fact that averaging several realizations of a random variable reduces the variance. The method was unsupervised and was adapted to denoise image sequences with an additive white noise while preserving the visual details on the movie frames. Very little attention has been paid to the problem of missing data (impulsive distortion) removal in image sequences. In [185] a 3D median filter for removing impulsive noise from image sequences was developed. This filter was implemented without motion compensation and so the results did not capture the full potential of these structures. Further, the median operation, although quite successful in the additive noise filtering in images, invariably introduces distortion when filtering of image sequences [185]. This distortion primarily takes the form of blurring fine image details.

The basic principles of despeckle filtering for still images presented in Chapter 2 of vol. I [1], i.e., the proposed despeckle filtering algorithms as well as the extraction of texture features, image quality evaluation metrics and the optical perception evaluation procedure by experts, can also be applied to video. The application of despeckle filters, the extraction of texture features, the calculation of image quality metrics, and the visual perception evaluation by experts may also be applied to video. The video can be broken into frames, which can then be processed one by one and then grouped together to form the processed video. Preliminary results for the application of despeckle filtering in ultrasound carotid and cardiac video were presented in this chapter. However, significant work still remains to be carried out.

In a work made by our group [3], we evaluated four different video despeckling filtering techniques (*DsFlsmv, DsFhmedian, DsFkuwahara* and *DsFsrad*) and applied them on 10 ultrasound videos of the CCA. Our effort was to achieve multiplicative noise reduction in order to increase visual perception by the experts but also to make the videos suitable for further analysis such as video segmentation and coding. The video despeckle results were evaluated based on visual perception evaluation by two experts, different texture descriptors and video quality metrics. The results showed that the best filtering method for ultrasound videos of the CCA is the *DsFlsmv* followed by the despeckle filter *DsFhmedian*. Both filters performed best with respect to the visual evaluation by the experts as well as by the video quality metrics. It is noted that the evaluation performance for the *DsFlsmv* was slightly better. Our results in [3] are also consistent with our previous despeckle filtering results found in other studies performed by our group [7, 8] on ultrasound images of the CCA, where the *DsFlsmv* filter was also found to be the preferred filter in terms of optical perception evaluation and classification accuracy between asymptomatic and symptomatic plaques.

While there are a number of despeckle filtering techniques proposed in the literature for despeckle filtering on ultrasound images of the CCA, we have found no other studies in the literate for despeckle filtering in ultrasound videos of the CCA. As it has been mentioned in the introduction a number of studies investigated additive noise filtering in natural video sequences [74, 75]. The usefulness of these methods in ultrasound video denoising of multiplicative noise still remains to be investigated.

Despeckle filtering is an important operation in the enhancement of ultrasonic video of the carotid artery. Initial findings show some promise of these techniques, however more work is needed to evaluate further the performance of the suggested despeckle filters. Future work will investigate the application of the proposed video despeckle filtering methods in a larger video dataset, as well as between asymptomatic and symptomatic patients in order to select the most appropriate filleting method for the two different classes. Furthermore, the proposed despeckle filtering techniques will be investigated and evaluated as a preprocessing step in CCA automated ultrasound video segmentation and in a mobile health telemedicine system.

There are several studies reported in the literature for filtering additive noise from natural video sequences [68]–[74], but we have found no other studies where despeckle filtering on ultrasound medical videos (of the CCA) was investigated. Previous research on the use of despeckle filtering of the CCA images [2, 7–14, 100, 147], and videos [13, 14] were also reported by our group where improved results were presented in terms of visual quality and classification accuracy between asymptomatic and symptomatic plaques. Moreover, it should be mentioned that a significant number of studies investigated different despeckle filters in various medical ultrasound video modalities with very promising results [11].

The performance of the proposed video despeckle filtering methods was evaluated in [14] after video normalization and despeckle filtering using visual perception evaluation, texture features, and image quality evaluation metrics.

The need for image standardization or post-processing has been suggested in the past, and normalization using only blood echogenicity as a reference point has been applied in ultrasound images of carotid artery [5]. Brightness adjustments of the ultrasound images and videos have been used in this book, as this has been shown to improve image compatibility, by reducing the variability introduced by different gain settings and facilitate ultrasound tissue comparability [5, 34, 174].

The images and videos used for the image texture analysis and quality evaluation were normalized manually by linearly adjusting the image so that the median gray level value of the blood was 0–5, and the median gray level of the adventitia (artery wall) was 180–190. The scale of the gray level of the images ranged from 0–255 [112]. The normalization can be made using the IDF [2] and VDF [3] toolboxes for image and videos, respectively. This normalization using blood and adventitia as reference points was necessary in order to extract comparable measurements in case of processing images obtained by different operators or different equipment [112]. The image normalization procedure was implemented in MATLAB™ software (version 6.1.0.450, release 12.1, May 2010, by The Mathworks, Inc.), and tested on a Pentium III desktop computer, running at 1.9 GHz, with 512 MB of RAM memory. The same software and computer station were also used for all other methods employed in this book.

3.6   EVALUATION OF TWO DIFFERENT ULTRASOUND SCANNERS BASED ON DESPECKLE FILTERING

In this section the results for the evaluation of image quality based on despeckle filtering performed on two different ultrasound imaging scanners (ATL HDI-3000 and ATL HDI-5000) are presented.

3.6.1   EVALUATION OF DESPECKLE FILTERING ON AN ULTRASOUND IMAGE

Figure 3.3 illustrate the original, NF, despeckled, DS, normalized, N, and normalized despeckled, NDS, images for the two ultrasound image scanners. It is shown that the images for the ATL HDI-3000 scanner have greater speckle noise compared to the ATL HDI-5000 images. Moreover, the lumen borders and the IMT are more easily identified with the ATL HDI-5000 on the N and NDS images.

3.6.2   EVALUATION OF DESPECKLE FILTERING ON GRAY-VALUE LINE PROFILES

Figure 3.2 shows gray-value line profiles, from top to bottom of an ultrasound carotid image (see Fig. 3.2a) for the original, NF, despeckled, DS, normalised, N, and normalised despeckled, NDS, images for the ATL HDI-3000 and ATL HDI-5000 scanner. Figure 3.3 also shows that speckle reduction filtering sharpens the edges. The contrast in the ATL HDI-3000 images was decreased after normalization and speckle reduction filtering, whereas the contrast for the ATL HDI-5000 images was increased after normalization.

3.6.3   EVALUATION OF DESPECKLE FILTERING BASED ON VISUAL PERCEPTION EVALUATION

Table 3.9 shows the results in percentage (%) format for the visual perception evaluation made by the two vascular experts on the two scanners. It is clearly shown that the highest scores are given for the NDS images, followed by the N, DS, and NF images for both scanners from both experts.

Table 3.10 presents the results of the Wilcoxon rank sum test for the visual perception evaluation, performed between the NF-DS, NF-N, NF-NDS, DS-N, DS-NDS, and N-NDS images, for the first and second observer on the ATL HDI-3000 and the ATL HDI-5000 scanner respectively. The results of the Wilcoxon rank sum test in Table 3.10 for the visual perception evaluation were mostly significantly different (S) showing large intra-observer and inter-observer variability for the different preprocessing procedures (NF-DS, NF-N, NF-NDS, DS-N, DS-NDS, N-NDS) for both scanners. Not significantly (NS) different values were obtained for both scanners, after normalization and speckle reduction filtering, showing that this improves the optical perception evaluation.

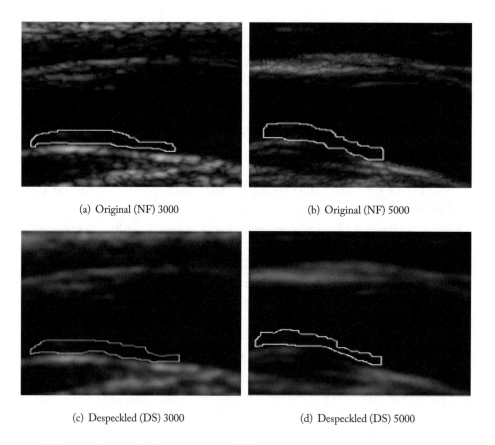

(a) Original (NF) 3000   (b) Original (NF) 5000

(c) Despeckled (DS) 3000   (d) Despeckled (DS) 5000

Figure 3.3: Ultrasound carotid plaque images of Type II outlined by an expert of the original (NF), speckle reduction (DS), normalized (N), and normalized speckle reduction (NDS), of the ATL HDI-3000, and ATL HDI-5000 shown in the left and right columns, respectively. Source [8], © MBEC 2006. *(Continues.)*

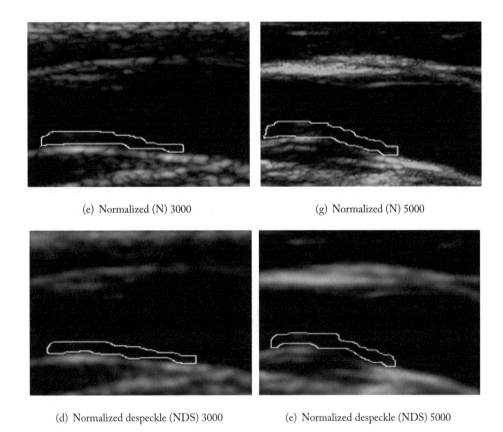

(e) Normalized (N) 3000   (g) Normalized (N) 5000

(d) Normalized despeckle (NDS) 3000   (e) Normalized despeckle (NDS) 5000

Figure 3.3: *(Continued.)* Ultrasound carotid plaque images of Type II outlined by an expert of the original (NF), speckle reduction (DS), normalized (N), and normalized speckle reduction (NDS), of the ATL HDI-3000, and ATL HDI-5000 shown in the left and right columns, respectively. Source [8], © MBEC 2006.

Table 3.9: Visual perception evaluation for the image quality on 80 images processed from each scanner for the original (NF), despeckled (DS), normalized (N), and normalized despeckled (NDS)

	Visual perception score							
Ultrasound scanner	ATL HDI-3000				ATL HDI-5000			
Pre-processing procedure	*NF*	*DS*	*N*	*NDS*	*NF*	*DS*	*N*	*NDS*
Angiologist	30	43	69	72	26	42	59	70
Neurovascular specialist	41	56	54	71	49	53	59	72
Average	36	50	62	72	38	48	59	71

Source [8], © MBEC 2006.

Table 3.10: Wilcoxon rank-sum test p value for the ATL HDI-3000 and the ATL HDI-5000 scanner for the visual perception evaluation performed by the experts between the NF-DS, NF-N, NF-NDS, DS-N, DS-NDS, and N-NDS images. The test shows in parenthesis with S significant difference at $p < 0.05$ and NS no significant difference at $p >= 0.05$

Ultrasound scanner	ATL HDI-3000					
Pre-processing procedure	*NF-DS*	*NF-N*	*NF-NDS*	*DS-N*	*DS-NDS*	*N-NDS*
Angiologist	$1.2*10^{-4}$	$1.1*10^{-11}$	$1.1*10^{-11}$	$1.3*10^{-8}$	$1.1*10^{-8}$	0.385
	(S)	(S)	(S)	(S)	(S)	(NS)
Neurovascular specialist	$2.9*10^{-4}$	0.004	$3.5*10^{-9}$	0.55	$1.7*10^{-4}$	$1.5*10^{-4}$
	(S)	(S)	(S)	(NS)	(S)	(S)
Ultrasound scanner	ATL HDI-5000					
Pre-processing procedure	*NF-DS*	*NF-N*	*NF-NDS*	*DS-N*	*DS-NDS*	*N-NDS*
Angiologist	0.14	0.001	$9.6*10^{-8}$	0.65	$8.9*10^{-6}$	$7.6*10^{-8}$
	(NS)	(S)	(S)	(NS)	(S)	(S)
Neurovascular specialist	0.85	$1.3*10^{-4}$	$6.1*10^{-8}$	0.56	0.002	0.001
	(NS)	(S)	(S)	(NS)	(S)	(S)

Source [8], © MBEC 2006.

3.6.4   EVALUATION OF DESPECKLE FILTERING BASED ON STATISTICAL AND TEXTURE FEATURES

Table 3.11 presents the results of the statistical and texture features for the 80 images recorded from each image scanner. The upper part of Table 3.11 shows that, the effect of speckle reduction filtering, DS, for both scanners was similar, that is the mean and the median were preserved, the standard deviation was reduced, the skewness and the kurtosis were reduced, and the speckle index was reduced (see also Fig. 3.1 and Fig. 3.2c, d, g, and h, where it is shown that the gray-value line profiles are smoother and less flattened). Furthermore, Table 3.11 shows that some statistical measures like the skewness, kurtosis, and speckle index, were better than the original, NF, and speckle reduction, DS, images after normalization, N, for both scanners, and were even better after normalization and speckle reduction, NDS. However, the mean was increased for N and NDS images for both scanners.

Table 3.11: Statistical and texture features (Mean values for 80 images processed from each scanner) for the original (NF), despeckled (DS), normalized (N) and normalized despeckled (NDS) images

Scanner	ATL HDI-3000				ATL HDI-5000			
Images	NF	DS	N	NDS	NF	DS	N	NDS
Statistical Features (SF)								
Mean	22.13	21.78	26.81	26.46	22.72	22.35	27.81	27.46
Median	3.07	4.53	3.56	5.07	3.73	5.23	4.59	6.07
Stand. Deviation (σ^2)	40.67	36.2	45.15	41.48	41.22	36.7	45.9	42.31
Skewness (σ^3)	2.88	2.49	2.23	2.00	2.84	2.45	2.17	1.94
Kurtosis (σ^4)	12.43	10.05	7.94	6.73	12.13	9.82	7.56	6.43
Speckle Index (C)	0.29	0.27	0.25	0.24	0.28	0.27	0.24	0.23
SGLDM-Range of Values								
Entropy	0.24	0.34	0.25	0.34	0.40	0.48	0.41	0.48
Contrast	667	309	664	303	618	302	595	287
ASM	0.36	0.35	0.38	0.37	0.37	0.33	0.39	0.35

SOURCE [8], © MBEC 2006.

In the bottom part of Table 3.12, it is shown that the entropy was increased and the contrast was reduced significantly in the cases of DS and NDS for both scanners. The entropy was slightly increased and the contrast was slightly reduced in the cases of N images for both scanners. The ASM was reduced for the DS images for both scanners and for the NDS images for the ATL HDI-5000 scanner.

Table 3.12 presents the results of the Wilcoxon rank sum test for the statistical and texture features (see Table 3.11), performed on the NF-DS, NF-N, NF-NDS, DS-N, DS-NDS, and N-NDS images on the ATL HDI-3000 scanner. No statistically significant difference was found in the first part of Table 3.12 when performing the non-parametric Wilcoxon rank sum test at $p < 0.05$, between the original, NF and despeckled, DS, the original, NF and normalized, N, and the original, NF and normalized despeckled, NDS features for both scanners. Statistical significant different values were mostly obtained for the second part of Table 3.12 for the ASM, contrast, and entropy.

Furthermore, Table 3.12 shows that, the entropy that is a measure of the information content of the image was higher for the ATL HDI-5000 in all the cases. The ASM that is a measure of

Table 3.12: Wilcoxon rank-sum test p value for the ATL HDI-3000 scanner for the statistical and texture features between the NF-DS, NF-N, NF-NDS, DS-N, DS-NDS, and N-NDS images. The test shows in parenthesis with S significant difference at $p < 0.05$ and NS no significant difference at $p >= 0.05$

	ATL HDI-3000					
Pre-processing procedure	*NF-DS*	*NF-N*	*NF-NDS*	*DS-N*	*DS-NDS*	*N-NDS*
	Statistical Features (SF)					
Mean	0.69	0.5	0.07	0.56	0.31	0.09
	(NS)	(NS)	(NS)	(NS)	(NS)	(NS)
Median	0.02	0.09	0.07	0.001	0.34	0.03
	(S)	(NS)	(NS)	(S)	(NS)	(S)
Stand. Deviation (σ^2)	0.01	0.02	0.08	0.03	0.004	$3.8*10^{-4}$
	(S)	(S)	(NS)	(S)	(S)	(S)
Skewness (σ^3)	0.08	0.45	$7.3*10^{-4}$	0.037	0.17	0.07
	(NS)	(NS)	(S)	(S)	(NS)	(NS)
Kurtosis (σ^4)	0.08	0.09	$4.5*10^{-4}$	0.19	0.34	0.07
	(NS)	(NS)	(S)	(NS)	(NS)	(S)
	SGLDM-Range of Values					
Entropy	$6.9*10^{-7}$	0.09	$2.2*10^{-3}$	$7.1*10^{-11}$	0.17	$4.2*10^{-5}$
	(S)	(NS)	(S)	(S)	(NS)	(S)
Contrast	$3*10^{-12}$	0.25	$4.2*10^{-7}$	$3.1*10^{-5}$	0.45	$5.6*10^{-9}$
	(S)	(NS)	(S)	(S)	(NS)	(S)
ASM	$9.6*10^{-7}$	$2.2*10^{-9}$	$1.4*10^{-6}$	$6.7*10^{-8}$	$7.2*10^{-7}$	$4.3*10^{-7}$
	(S)	(S)	(S)	(S)	(S)	(S)

SOURCE [8], © MBEC 2006.

the inhomogeneity of the image is lower for the ATL HDI-5000 in the cases of the DS and NDS images. Furthermore, the entropy, and the ASM were more influenced from speckle reduction than normalization as, they are reaching their best values after speckle reduction filtering.

3.6.5   EVALUATION OF DESPECKLE FILTERING BASED ON IMAGE QUALITY EVALUATION METRICS

Table 3.13 illustrates the image quality evaluation metrics, for the 80-ultrasound images recorded from each image scanner, between the NF-DS, NF-N, NF-NDS, and N-NDS images. Best values were obtained for the NF-N images with lower RMSE, Err3, and Err4, higher SNR, and PSNR for both scanners. The GAE was 0.00 for all cases, and this can be attributed to the fact that the information between the original and the processed images remains unchanged. Best values for Q and SSIN were obtained for the NF-N images for both scanners, whereas best values for SNR were obtained for the ATL HDI-3000 scanner on the NF-N images.

Table 3.13: Image quality evaluation metrics between the original-despeckled (NF-DS), original-normalized (NF-N), original-normalized despeckled (NF-NDS) and the normalized-normalized despeckled (N-NDS) images

Evaluation metrics	ATL HDI-3000				ATL HDI-5000			
	NF-DS	NF-N	NF-NDS	N-NDS	NF-DS	NF-N	NF-NDS	N-NDS
MSE	1.4	1.3	2.0	1.3	1.2	0.3	1.9	1.3
RMSE	1.2	0.4	1.4	1.1	1.1	0.5	1.3	1.1
Err 3	3.8	0.8	3.9	3.5	3.7	0.8	3.8	3.5
Err 4	8.2	1.2	8.0	7.5	8.1	1.3	7.8	7.5
GAE	0	0	0	0	0	0	0	0
SNR	5.0	16.5	4.8	5.4	5.3	15.9	5.1	5.4
PSNR	48.0	59	45.6	44.6	47.4	58.5	46	44.6
Q	0.7	0.93	0.73	0.69	0.72	0.93	0.72	0.71
SSIN	0.9	0.95	0.92	0.83	0.94	0.95	0.91	0.83

Source [8], © MBEC 2006.

Table 3.13 shows that the effect of speckle reduction filtering was more obvious on the ATL HDI-3000 scanner, which shows that the ATL HDI-5000 scanner produces images with lower noise and distortion. Moreover, it was obvious that all quality metrics presented here are equally important for image quality evaluation. Specifically, for the most of the quality metrics, better measures were obtained between the NF-N, followed by the NF-NDS, and N-NDS images for both scanners. It is furthermore important to note that a higher PSNR (or equivalently, a lower RMSE) does not necessarily imply a higher subjective image quality, although they do provide some measure of relative quality.

Furthermore, the two experts evaluated visually 10 B-mode ultrasound images with different types of plaque [5] (see Fig. 3.3), by delineating the plaque at the far wall of the carotid artery wall. The visual perception evaluation, and the delineations made by the two experts, showed that the plaque may be better identified on the ATL HDI-5000 scanner after normalization and speckle reduction, NDS, whereas the borders of the plaque and the surrounding tissue may be better visualized on the ATL HDI-5000 when compared with the ATL HDI-3000 scanner.

Table 3.14 summarizes the image quality evaluation results of this study, for the visual evaluation (Table 3.9), the statistical and texture analysis (Table 3.11), and the image quality evaluation metrics (Table 3.13). A double plus sign in Table 3.14 indicates very good performance, while a single plus sign a good performance. Table 3.14 can be summarized as follows: i) the NDS images were rated visually better on both scanners, ii) the NDS images showed better statistical and texture analysis results on both scanners, iii) the NF-N images on both scanners showed better image quality evaluation results, followed by the NF-DS on the ATL HDI-5000 scanner and the NF-DS on the HDI ATL-3000 scanner, iv) the ATL HDI-5000 scanner images have considerable higher entropy than the ATL HDI-3000 and thus more information content. However, based on the visual evaluation by the two experts, both scanners were rated similarly.

Table 3.14: Summary findings of Image quality evaluation in ultrasound imaging of the carotid artery

Ultrasound scanner	Visual evaluation Table 3.9				Statistical and texture analysis Tables 3.11-3.12				Image quality evaluation Table 3.13			
	NF	DS	N	NDS	NF	DS	N	NDS	NF-DS	NF-N	NF-NDS	N-NDS
ATL HDI-3000				++		+		++		++	+	
ATL HDI-5000				++		+		++	+	++		

Source [8], © MBEC 2006.

CHAPTER 4

Wireless Video Communication Using Despeckle Filtering and HEVC

Andreas Panayides, *University of Cyprus*

In this chapter we discuss the use of despeckle filtering prior to video coding for mobile-health (mHealth) medical video communications. Linked with the new High Efficiency Video Coding (HEVC) standard, the use of despeckle filtering facilitates significant bitrate gains for equivalent clinical quality, which can provide for beyond high-definition (HD) real-time communication of ultrasound video over emerging wireless networks. The presented study is based on [188–190].

4.1 MOBILE HEALTH MEDICAL VIDEO COMMUNICATION SYSTEMS: INTRODUCTION AND ENABLING TECHNOLOGIES

Medical video communication is a key bandwidth-demanding component of mHealth applications ranging from emergency incidents response, to home monitoring, and medical education [188]–[193]. In-ambulance video (trauma and ultrasound) communication for remote diagnosis and care can provide significant time savings which can prove decisive for the patient's survival. Similarly, emergency scenery video can assist in better triage and preparatory hospital processes. Remote diagnosis allows access to specialized care for people residing in remote areas, but also for the elderly, and people with chronic diseases and mobility problems. Moreover, it can support mass population screening and second opinion provision, especially in developing countries. Medical education also benefits from real-time surgery video transmission as well as ultrasound examinations.

Overall, wireless medical video communication poses significant challenges that stem from limited bandwidths over noisy channels. In terms of both bandwidth and processing requirements, medical videos dominate over other biomedical signals. Clearly, the evolution of future mHealth systems will depend and also benefit from the development of effective medical video communication systems.

MHealth medical video communication systems' impressive growth over the past decade is primarily attribute to associated advances in video compression and wireless network technologies. The former allow real-time, robust and efficient encoding, while recent video coding standards also encompass a network abstraction layer for higher flexibility. The latter, facilitate constantly increasing data transfer rates, extended coverage, reduced latencies and transmission reliability. In what follows, we highlight video coding standards and wireless transmission technologies evolution timeframe.

4.1.1   VIDEO COMPRESSION TECHNOLOGIES

In terms of video coding standards, efficient video compression systems were introduced over the last decades. In early 1990's the international telecommunication union-telecommunication sector (ITU-T) developed the first, H.261 [194] video coding standard, which was originally designed for video-telephony and videoconferencing applications. H.261 supported the quarter common intermediate format (QCIF-176x144) and the common intermediate format (CIF-352x288) video resolutions. Subsequently, H.262 [195] released in 1995 facilitated interlaced video provision and increased video resolutions support to 4CIF (720x576) and 16CIF (1408x1152). Its successor, termed H.263 [196] introduced in 1996 provided for improved quality at lower bit rates and also allowed lower, sub-QCIF (128×96) video resolution encoding. The highly successful H.264/AVC [197] standard was released by ITU-T in 2003 and accounted for bit rate demands reductions of up to 50% for equivalent perceptual quality compared to its predecessor [198]. The current state-of-the-art video coding standard is the High Efficiency Video Coding (HEVC) standard [199], standardized in 2013. HEVC supports video resolutions ranging from 128×96 to 8192×4320 and provides 50% bit rate gains for comparable visual quality compared to H.264/AVC [200].

H.264/AVC

H.264/AVC was jointly developed by the ISO/IEC motion pictures experts group (MPEG) and ITU-T video quality experts group (VCEG) who formed the joint video team (JVT). H.264/AVC met the growing demand for multimedia and video services by providing enhanced compression efficiency significantly outperforming all prior standards. H.264/AVC comprises of numerous advances in standard video coding technology, coding efficiency improvement, error robustness enhancement, and flexibility for effective use over a variety of network types and application domains [198]. H.264/AVC defines a video coding layer (VCL) and a network abstraction layer (NAL) to enable transportation over heterogeneous networks. VLC is responsible for video coding, maintaining the block-based motion-compensated hybrid video coding concept. Compared to prior standards, VCL provides enhanced entropy coding methods, uses a small block-size exact-match transform, facilitates adaptive in-loop deblocking filter and enhances motion prediction capability. On the other hand, NAL is a novel concept aiming at a network-friendly adaptation of VCL content to candidate heterogeneous networks and/or storage devices (or cloud).

H.264/AVC defines different profiles and levels. Each profile and level specify restrictions on bit streams, hence limits on the capabilities needed to decode a bit stream [197]. Baseline, main, extended and high profiles assume different processing devices tailored for different applications and provide incremental level capabilities (and therefore complexity).

H.264/AVC facilitates a broad range of error resilience techniques for a wide variety of applications. Toward this end, one of H.264/AVC key error resilience features is flexible macroblock ordering (FMO) which defines macroblock transmission order to allow for easier recovery. Using this feature, each frame may be partitioned in up to several different slices. This feature also allows region-of-interest (ROI) coding and recovery (a scheme extensively used in mHealth medical video communication systems, see Section 4.2) or arbitrary spatial placement of blocks for better concealment during interpolation. Another important feature is redundant slices (RS) where a slice is redundantly inserted in the communicated bits ream to maximize the video's error resilience.

4.1.2 HIGH EFFICIENCY VIDEO CODING (HEVC)

On April 2013 HEVC was officially approved as an ITU standard and on June 2013 [199] was formally published on the ITU-T website. HEVC is also known as H.265 and MPEG-H Part 2 and was developed by the joint collaborative team on video coding (JCT-VC).

HEVC introduces new coding tools as well as significant improvements of components already known from H.264/AVC. New tools in HEVC include variable size block partitioning using quadtrees for the purpose of prediction and transformation and an additional in-loop filter, namely sample adaptive offset (SAO). Improvements include additional intra-prediction angles, advanced motion vector prediction (AMVP), a new block merging mode that enables neighboring blocks to share the same motion information, larger transform sizes and a more efficient transform coefficient coding [200]. HEVC incorporates only one entropy coder, context adaptive binary arithmetic coding (CABAC) borrowed from H.264/AVC.

One of the key differences of HEVC compared to H.264/AVC is the frame coding structure. In HEVC each frame is partitioned into coding tree blocks (CTBs) [201]. The luma CTB and the two chroma CTBs, together with the associated syntax, form a coding tree unit (CTU). The CTU is the basic processing unit of the standard to specify the decoding process and replaces the macroblock structure found in all prior video coding standards. A CTB may contain a single coding unit (CU) or can be split to form multiple CUs [200]. CUs can be further split into prediction units (PUs) used for intra- and inter-prediction and transform units (TUs) defined for transform and quantization. As the picture resolution of videos increases from standard definition to high definition (HD) and beyond, the chances are that the picture will contain larger smooth regions, which can be encoded more effectively when large block sizes are used. Based on the afore-mentioned assumption HEVC supports up to 64 × 64 pixels encoding blocks compared to H.264/AVC that only supports up to 16 × 16 pixels blocks.

HEVC is the first video coding standard to incorporate features that provide for parallel processing [200]. The new tiles tool allows the partitioning of a picture into independently decoded rectangular segments of approximately equal CTUs. Tiles increase flexibility compared to normal slices in H.264/AVC and incorporate considerably lower complexity than FMO. Wave front parallel processing (WPP) splits slices into rows of CTUs, which are then processed in parallel provided a certain time window has advanced since the processing of the immediately prior row to allow time for decisions relating to entropy coding. Dependent slices allow a slice using tiles or WPP coding tools to be fragmented and associated with different NAL packets. Dependent slices are associated with lower encoding time but a slice may be decoded provided part of the decoding process of the dependable slice has been performed.

4.2   WIRELESS INFRASTRUCTURE

The global system for mobile communications (GSM) signified the transition from analog 1st generation (1G) to digital 2nd generation (2G) technology of mobile cellular networks. In the past two decades, mobile telecommunication networks are continuously evolving. Milestone advances range from 2.5G (general packet radio service (GPRS) and enhanced data rates for GSM evolution (EDGE)) and 3G (universal mobile telecommunications system (UMTS)) wireless networks, to 3.5G (high speed downlink packet access (HSDPA), high speed uplink packet access (HSUPA), high speed packet access (HSPA), and HSPA+), mobile WiMAX networks, and long term evolution (LTE) systems. The afore-described wireless networks facilitate incremental data transfer rates while minimizing end-to-end delay. In other words, features in wireless communications follow those of wired infrastructure with a reasonable time gap of few years. The latter enables the development of responsive mHealth systems suitable for emergency telemedicine [191]. Evolving wireless communications networks' theoretical upload data rates range from 50 kbps–86 Mbps, and extend up to 100 Mbps for LTE-Advanced and WirelessMan-Advanced 4G systems. In practice, typical upload data rates are significantly lower. More specifically typical upload data rates range from (i) GPRS: 30–50 kbps, (ii) EDGE: 80–160 kbps, (iii) evolved EDGE: 150–300 kbps, (iv) UMTS: 200–300 kbps, (v) HSPA: 500 kbps–2 Mbps, (vi) HSPA+: 1–4 Mbps, and (vii) LTE: 6–13 Mbps [202] (subject to the implemented Release and deployment specific details).

4.2.1   4G NETWORKS CONFIRMING TO IMT-ADVANCED REQUIREMENTS

WirelessMAN-Advanced and LTE-advanced technologies based on IEEE 802.16m and 3GPP Release 10 specifications respectively, satisfy the international mobile telecommunications-advanced (IMT-advanced) requirements specified by the international telecommunication union-radio communication sector (ITU-R) [203], and are officially considered as 4G technologies.

Being backwards compatible, this family of technologies targets improved uplink and downlink rates of 100 Mbps and 1 Gbps respectively, increased coverage and throughput, enhanced mobility support, reduced latencies, enhanced quality of service (QoS) provision, efficient spectrum usability and bandwidth scalability, and security, with simple architectures, in favor of the end user.

4.2.2   WORLDWIDE INTEROPERABILITY FOR MICROWAVE ACCESS (WIMAX)

Worldwide Interoperability for Microwave Access (WiMAX) was firstly standardized for fixed wireless applications in 2004 by the IEEE 802.16-2004d and then for mobile applications in 2005 by the IEEE 802.16e standards. Current standardization 802.16m [204], also termed as IEEE WirelessMAN-Advanced, aims to achieve higher market penetration.

4G WiMAX frequency bands facilitate deployment between 450–3600 MHz [204]. Channel bandwidth allows great flexibility in the sense that it allows WiMAX operators to consider channel bandwidths between 1.25, 2.5, 5, 10, and 20 MHz (802.16e). In 802.16m scalable bandwidth between 5–40 MHz for a single RF carrier is considered, extended to 100 MHz with carrier aggregation. WiMAX employs a set of high and low level technologies to provide for robust performance in both line-of-sight (LOS) and non-line-of-site (NLOS) conditions.

PHY layer's central features include adaptive modulation and coding, hybrid automatic repeat request (HARQ), and fast channel feedback. Key technology in the success of WiMAX systems in general is the orthogonal frequency division multiplexing (OFDM) scheme employed in the PHY layer. OFDM, and more specifically scalable orthogonal frequency division multiple access (SOFDMA), allows dividing transmission bandwidth into multiple subcarriers. The number of subcarriers starts from 128 for 1.25 MHz channel bandwidth and extends up to 2048 for 20 MHz channels. In this manner, dynamic QoS tailored to individual application's requirements can be succeeded. In other words, multipath interference is addressed by employing OFDM while available bandwidth can be split and assigned to several requested parallel applications for improved system's efficiency. The latter is true for both downlink (DL) and uplink (UL). Multiple input multiple output (MIMO) antenna system allows transmitting and receiving multiple signals over the same frequency.

In the MAC layer, the most important supported features can be summarized in QoS provision through different prioritization classes, direct scheduling for DL and UL, efficient mobility management, as well as security. The 5 QoS categories facilitated by WiMAX networks are discussed in [204]. For real-time video streaming, as in the case of emergency telemedicine scenarios, real-time Polling Service (rtPS) QoS class best suits the applications requirements. Mobility management is also well addressed in 802.16e and current 802.16m standards, providing mobility support of up to 120 km/h and 350 km/h, respectively. Enhanced security is based on extensible authentication protocol and advanced encryption.

A thorough overview of WiMAX standardization process, evolving concepts, technologies and performance evaluation of IEEE 802.16m appears in [204]–[206]. Key features of physical (PHY) and medium access control (MAC) layers are discussed next.

4.2.3   LONG TERM EVOLUTION (LTE)

LTE facilitates significant improvements with respect to 3G and HSPA systems. It provides increased data rates, improved spectral efficiency and bandwidth flexibility ranging between 1.4–

20 MHz, reduced latency and seamless handover. While being backwards compatible enabling seamless deployment on existing infrastructure, LTE shares a set of new cutting edge technologies and simple architecture.

In the physical layer, multiple-carrier multiplexing OFDMA is adopted for the downlink, while single carrier frequency division multiple access (SC-FDMA) is the access scheme used in the uplink. SC-FDMA utilizes single carrier modulation, orthogonal frequency multiplexing, frequency domain equalization, and has similar performance to OFDMA. Frequency division duplex (FDD) and time division duplex (TDD) are jointly supported in a single radio carrier. LTE allows multi-antenna applications for single and multi-users through MIMO technology (up to 4-layers in the downlink and 2-layers in the uplink), and supports different modulation and coding schemes. Automatic repeat request (ARQ) and hybrid-ARQ are implemented for increased robustness in data transmission.

Compared to LTE, LTE-advanced systems in Release 10 define further improvements in peak data rate, spectrum efficiency, throughput and coverage, as well as latency reductions are facilitated. Data rates in the order of 1 Gbps for low mobility and 100 Mbps for high mobility are achieved via the adoption of new technologies such as carrier aggregation, which enables wider bandwidth transmission up to 100 MHz. Enhanced MIMO techniques in LTE-advanced systems include 8-layer transmission in the downlink and 4-layer transmission in the uplink. Coordinated multipoint transmission and reception (CoMP) is another novel technique defined in the standard which provides for increased throughput on the cell edge. Relaying techniques for efficient cell deployment and coverage, and parallel processing for even greater reductions in latency are also exploited.

A comprehensive review of LTE and LTE-advanced technologies, comparative analysis, utilization scenarios combining different technology components to demonstrate conformance to IMT-advanced requirements, and also IMT-advanced evaluation results can be found in [207, 208].

4.3   SELECTED MHEALTH MEDICAL VIDEO COMMUNICATION SYSTEMS

In this section we review mHealth medical video communication systems' approaches spanning over the last decade (see Table 4.1). We focus on recent, diagnostically relevant approaches and discuss state-of-the-art methods. Compared to standard approaches in wireless video communications, m-health systems need to be diagnostically driven. This notion is derived from the objective of delivering medical video of adequate diagnostic quality. The latter differs from a focus on perceptual quality of conventional video, often termed subjective quality. Clinical quality cannot be compromised. Further to that, appropriate clinical protocols need to be established so as to guarantee that the communicated video evaluated by the remote medical expert is of the same diagnostic quality as the one displayed on the in-hospital machine screen. As a result, incorporated methods for video compression, wireless transmission, and clinical video quality as-

Table 4.1: Selected mHealth Medical Video Communication Systems. Source [189] © *Springer Series in Bio–/Neuroinformatics*, 2015. (*Continues.*)

	Author	Year	Resolution, Frame Rate, BitRate	Encoding Standard	Wireless Network	Medical Video Modality
Non-Diagnostically Driven Systems	Chu et al. [220][2]	04	{320x240 and 160x120} <5fps 50-80 Kbps	M-JPEG	3G-CDMA	Trauma video
	Garawi et al. [221][2,3]	06	176x144 @ 5fps 18.5-60 Kbps	H.263	3G-UMTS	Cardiac Ultrasound
	Alinejad et al. [217][1,2]	12	{176x144, 352x288} @ 10/20 fps {220, 430} Kbps, 1.3 Mbps	Windows Media Video (WMV)	3.5G - Mobile WiMAX, HSDPA	Cardiac Ultrasound
	Istepanian et al. [222][2,3]	09	176x144 @ 8-10fps 50-130 Kbps	H.264/ AVC	3.5G –HSDPA	Abdomen Ultrasound
	Pedersen et al. [223][2,3]	09	320x240 @ 10fps 349 Kbps	H.264/ AVC (Scalable)	WLAN and 3.5G - HSPA	Echocardiogram
	Hewage et al. [224][1,3]	11	NA, 50fps, NA	H.264/AVC (MVC)	LTE	Endoscopy Procedures
	Panayides et al. [233][2,3]	13	{176x144, 352x288, 560x416} @ 15fps, 64 – 768 Kbps	H.264/AVC	3.5G - HSPA	Carotid Artery Ultrasound
	Panayides et al. [189][1,3]	13	560x416 @ 40fps, up to 2 Mbps	H.264/AVC and HEVC	3.5G and beyond	Carotid Artery Ultrasound

Table 4.1: *(Continued.)* Selected mHealth Medical Video Communication Systems. Source [189] © *Springer Series in Bio-Neuroinformatics*, 2015.

	Year		Codec	Network	Application
Rao et al.[209][1,3,4]	09	360x240 @ 30 fps 500 Kbps	MPEG-2	3G and beyond	Paediatric respiratory distress related videos
Tsapatsoulis et al. [210][1,3,4]	07	352x288 @ 10fps 10 videos average: 507.2 Kbps	MPEG-2 and MPEG-4	3G and beyond	Carotid Artery Ultrasound
Martini et al. [211][1,4]	10	480x256@15fps 300 Kbps	H.264/AVC	3.5G - Mobile WiMAX	Cardiac Ultrasound
Panayides et al. [212][1,3,4]	11	352x288 @ 15fps 197-421 Kbps	H.264/AVC	3G and beyond	Carotid Artery Ultrasound
Panayides et al. [213][1,3,4]	12	{176x144, 352x288, 640x480} @15/ 30fps 200 Kbps -1.1 Mbps	H.264/AVC	3G and beyond	Abdominal Aortic Aneurysm
Debono et al. [218][1,4]	12	640x480 @ 25fps	H.264/AVC	3.5G - Mobile WiMAX	Cardiac Ultrasound
Khire et al. [215][1,3,4]	12	720x480 @ 30fps, 125 – 200 Kb/s	H.264/AVC	3G and beyond	Maxillofacial Surgery Clips
Panayides et al. [214][1,3,4]	13	704x576@15fps 768 Kb/s -1.5 Mb/s	H.264/AVC	3.5G - Mobile WiMAX	Carotid Artery Ultrasound
Cavero et al. [1,3][216]	13	720x576 @ 25fps 40 Kb/s (M-mode), 200 Kb/s (B-mode)	SPIHT	3G and beyond	Cardiac Ultrasound
Cavero et al. [217][1,3]	12	720x576 @ 25fps, 200 Kb/s	SPIHT	3.5G - HSUPA, mobile WiMAX	Cardiac Ultrasound

Diagnostically Driven Systems

[1]Simulation, [2]Real-Time, [3]Clinical Evaluation, [4]d-ROI

sessment (c-VQA) are developed exploiting the clinical aspect of the underlying medical video modality. The aim is to maximize the clinical capacity of the communicated video. Toward this end, diagnostically driven approaches are not always universally applicable; rather they are often medical video modality specific.

Figure 4.1 summarizes a diagnostically driven medical video communication framework. Following pre-processing steps, diagnostically relevant encoding takes place. Wireless transmission prioritizes the communicated video bit stream compared to less demanding applications. At the receiver's side, post-processing and diagnostically relevant decoding precedes video quality assessment. Where applicable, adaptation to the varying wireless network state is performed, using QoS and c-VQA measurements, to preserve the communicated video's clinical standards. The latter approach, excluding the wireless network component is used for fine tuning diagnostically acceptable source encoding parameters, including video resolution, frame rate, and quantization factor.

4.3.1   DIAGNOSTICALLY DRIVEN MHEALTH SYSTEMS

Diagnostically driven systems exploit the properties of the underlying medical video modality aiming to maximize the clinical capacity of the communicated medical video. They range from diagnostically relevant and resilient encoding, to reliable wireless communications based on the communicated video's clinical significance, and clinical video quality assessment methods.

4.3.2   DIAGNOSTIC REGION(S)-OF-INTEREST

Each medical video modality is characterized by unique properties which are assessed by the relevant medical expert during an examination. A diagnosis is provided based on a clinically established protocol that considers different clinical criteria. These criteria often relate to specific video regions, as for example in routine ultrasound screening examinations. In other words, specific video regions carry the clinical information required by the medical expert to assess the patient's status during a visit. The latter observation has been the driving force of diagnostically relevant approaches using diagnostic regions-of-interest (d-ROI).

Diagnostic ROI outline the clinically important video regions. Essentially, there exist two ways of defining d-ROI: a) using modality aware segmentation algorithms, and b) denoted by the medical expert. Both approaches can be established in standard clinical practice, and the selected mode is application specific and depends on resources availability. MHealth medical video communication systems using d-ROI dominate over other diagnostically driven approaches [188]. They have become the standard method to use in a very short time since they were first proposed. Besides the obvious efficiency that such systems introduce, the latter is also attributed to the plethora of different approaches and remarkable flexibility that these schemes enjoy.

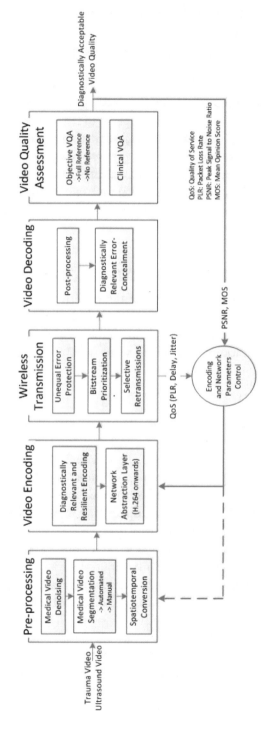

Figure 4.1: Medical video communication system diagram. Following pre-processing, diagnostically driven video encoding adapts to each medical video modality. Wireless transmission protects more strongly the clinically sensitive regions. At the receiver's side, diagnostically relevant error concealment is performed, followed by objective and clinical VQA. Cross-layer information is used to adapt to the underlying wireless network's varying state. Source [188] © *IEEE Signal Process. Mag.*, 2013.

4.3.3   DIAGNOSTICALLY RELEVANT ENCODING

Variable quality slice encoding based on the video region's diagnostic significance is a popular diagnostically driven method that builds on d-ROI. This scheme assigns quality levels according to the clinical significance of each d-ROI. Higher quality (less compression, more bits) is assigned to the most sensitive—clinically wise- regions while lower quality (higher compression, less bits) to the background, non-diagnostically important regions. The effectiveness of this approach has been demonstrated in [209]–[215] where significant bitrate demands reductions were documented without compromising perceptual quality. In earlier studies based on MPEG-2 and MPEG-4 video coding standards [209, 210] smoothing filters have been applied on the non-diagnostically important regions. On the other hand, more recent studies based on H.264/AVC make use of a customized version of the FMO error resilient technique [211]–[214], or use a similar concept for macroblock-level variable quality slice encoding [215]. Diagnostically relevant encoding approaches based on echocardiogram's mode properties which also lower bitrate requirements but without considering d-ROI is presented in [216, 217].

4.3.4   DIAGNOSTICALLY RESILIENT ENCODING

Diagnostically resilient encoding/decoding adapts error resilience/concealment techniques to the d-ROI. Error resilience techniques such as intra updating (frame level) can be tailored to match the periodicity of the cardiac cycle. Error-free cardiac cycles assist in communicating diagnostically lossless medical video. Intra updating (MB level prior to HEVC, CTU onwards) can match the d-ROI so that only the clinically sensitive regions are intra updated at specified intervals. Moreover, redundant slices of only the diagnostically important regions can be inserted in the communicated bit stream for maximizing error resilience [212]–[214]. Similarly, sophisticated error concealment or post processing techniques can be applied on the d-ROI at the receiver's end [218]. Afore-described schemes emphasize on enhancing/preserving the diagnostic capacity of the communicated medical video, while efficiently tackling the available resources.

4.3.5   RELIABLE WIRELESS COMMUNICATION

Reliable wireless communication is of paramount importance in mHealth systems. The quality of the transmitted medical video cannot be compromised by the error-prone nature and varying state of wireless channels. Therefore, the appropriate mechanisms should be installed so that adaptation to the current wireless network state is performed. The latter involves both a priori adaptation models, as well as real-time adaptation methods. Diagnostically relevant procedures include unequal error protection (UEP) schemes for the clinically sensitive regions such as forward error protection (FEC) codes [211, 217]. Where applicable, retransmissions of data packets conveying d-ROI information can be also adopted. This is the case for 3.5G and beyond wireless networks provided low end-to-end delay for ARQ and HARQ schemes, yet this scheme's efficiency remains to be investigated. Toward this end, MAC layer service prioritization is now a standard feature in 3.5G systems, enabling key video streaming applications -such mHealth systems- to receive traffic prioritization, in addition to guaranteed bitrate and low latencies. This service allows for a priori adaptation to the wireless channel's specification and has become standard in medical video communication [211, 214, 218, 219].

Cross layer approaches are widely used in the literature for adapting to the wireless channel's varying state [217]–[221]. Real-time monitoring of wireless network's QoS parameters allows a decision algorithm to trigger a switch to a pre-constructed (encoder) state that preserves the desired QoS threshold values. These threshold values that correspond to medical video of acceptable diagnostic quality are determined a priori. A switch criterion is usually based on a weighted cost function that includes a fusion of source encoding and network parameters, objective VQA measurements (e.g., Peal Signal to Noise Ratio (PSNR)), and c-VQA (mean opinion score (MOS) of clinical ratings provided by the relevant medical experts).

4.3.6   CLINICAL VIDEO QUALITY ASSESSMENT

Clinical VQA validates medical video communication systems objective of delivering medical video of adequate diagnostic quality over the wireless medium to a remote location. It largely differs from conventional perceptual quality evaluation, termed subjective quality. While low rate transmission errors may be generally tolerable in video streaming applications, in medical video communication, these errors cannot be allowed to compromise the diagnostic capacity of the transmitted medical video. At the remote physician's site the reconstructed medical video has to reproduce the quality of the in-hospital examination.

Central to the success of mHealth medical video communication systems is to a priori establish the range of diagnostically acceptable source encoding parameters. This is an essential pre-processing step which was primarily attributed to the inability of wireless networks to support medical video communication at the clinically acquired setup. The latter is expected to fade in the near future once 4G systems are widely deployed. Still, the minimum acceptable compression levels have to be clinically established by the relevant medical expert, given the variations between different medical video modalities. This procedure is a major component of the clinical evaluation procedure. Acceptable video resolutions that do not compromise the geometry characteristics, frame rates that preserve clinical motion, and compression ratios (both in terms of measured PSNR and quantization parameters) that maintain diagnostically acceptable quality levels are shown for different medical video modalities in [209, 212, 213, 216, 221, 222].

Presently, objective VQA algorithms do not correlate with medical experts mean opinion scores. As a result, there is a need to design new, diagnostically driven c-VQA metrics that will be eventually used to predict clinical scores. Clinical VQA is based on clinically established protocols that assess different clinical criteria. As already described, these clinical criteria often relate to specific d-ROI. This property can be exploited in future c-VQA metrics that will provide a weighted output function based on the region's clinical significance. The latter is briefly highlighted in [212], where VQA measurements performed over the primary d-ROI correlated better to the clinical ratings. The most detailed protocol for echocardiogram evaluation using c-VQA principles appears in [223]. Physiological properties -such as the periodicity of the cardiac cycle- are key features that relate to diagnostic capacity as highlighted earlier. Consecutive error-free cardiac cycles allow the medical expert to reach a confident diagnosis. This is an interesting research subject that has not been adequately addressed in the current literature.

4.4 ULTRASOUND VIDEO COMMUNICATION USING DESPECKLE FILTERING AND HEVC

The basic system diagram is presented in Fig. 4.2. Despeckle filtering is applied prior to video compression to improve quality while reducing bandwidth requirements. The filtered image is then encoded using a variety of different standards and then decoded in order to allow for comparisons with the original video. Validation of the system includes objective VQA, clinical evaluation based on mean opinion scores (MOS), and methods comparison to determine bitrate gains. We provide details on each component below.

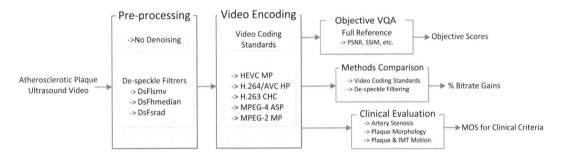

Figure 4.2: Ultrasound Video Encoding & Evaluation System Diagram. Ultrasound video de-noising precedes video encoding. The user can select the appropriate despeckle filtering algorithm and the most efficient video compression standard. Video quality assessment includes clinical evaluation by the relevant medical expert as well as objective measurements. De-speckle filtering methods and video coding standards comparison provides the bitrate gains achieved by the best performing methods. Source [214] © *IEEE J. Biomed. Health Inform*, 2013.

4.4.1 METHODOLOGY

The despeckle filters used for pre-processing were the *DsFlsmv*, *DsFhmedian* and *DsFsrad* which were presented in volume I of this book [1]. These filters were selected as they have achieved the best performance in terms of visual (clinical) quality as assessed by medical experts, edge and texture preservation, and image quality evaluation performance.

4.4.2   VIDEO CODING STANDARDS COMPARISON

The highest efficiency modes are selected for each encoding standard as given in [200]. Indicatively, for HEVC, all new coding tools are enabled–except weighted prediction–, as per the single defined profile in the standard termed Main Profile (HEVC MP). For H.264/AVC the High Profile was selected (H.264/AVC HP), also with weighted prediction disabled. For MPEG-4 the Advanced Simple Profile (MPEG-4 ASP) was used, while the Conversational High Compression was selected for H.263 (H.263 CHC), and lastly the Main Profile was used for MPEG-2/H.262 (MPEG-2MP). For a fair comparison, we vary the quantization parameters to achieve a similar range of rate-distortion performance for all standards. Furthermore, the quantization parameter step size is selected so that a single step results in PSNR increase of approximately 3 dB. More specifically, for MPEG-2, MPEG-4, and H.263, ultrasound videos are encoded using QPs ranging from 2 to 31 using QP = 2, 3, 4, 5, 6, 8, 10, 13, 16, 20, 25, 31, while for H.264/AVC and HEVC QPs range between 20 and 42, with a QP step size of two. For all cases, 200 frames per video sequence were encoded and an intra encoded frame (I-frame) was inserted every 48 frames.

4.4.3   VIDEO QUALITY ASSESSMENT

We consider the PSNR and the SSIM [224] for assessing video quality. Here, we note that the average PSNR and SSIM are computed over each video frame [225] and then averaged over the entire video. Both the PSNR and the SSIM are full-reference methods that require access to the original, uncompressed videos. For evaluating image quality, SSIM correlates significantly better to perceived, visual quality than the standard use of PSNR [226]. However, it does not assess the motion of the reconstructed videos as required in our application. The subject of video quality assessment is still an open area of research. For our application, we use extensive clinical VQA methods to properly address these issues.

4.4.4   RATE-DISTORTION COMPARISONS

To estimate bitrate savings, we compute the percentage savings for equivalent (objective) video quality. This is accomplished using the BD-rate algorithm [227]. The BD-rate algorithm is used to compute the objective differences between two rate-distortion curves and provides the percentage bitrate difference. The rate-distortion curves for the compared methods are constructed as functions of twelve rate points based on the luma PSNR (Y-PSNR). The final percentage difference is averaged over the examined data set.

4.4.5   CLINICAL VIDEO QUALITY ASSESSMENT

Clinical video quality assessment aims to address the diagnostic-quality of the reconstructed videos. There are three diagnostic region(s) of interest that are considered here:

1. Atherosclerotic plaque region: This is the primary d-ROI (see caption in Fig. 4.3) and it is used to determine the plaque's type by assessing the plaque's morphology and texture characteristics.

2. Near and far wall regions: Visualizing the artery walls and associated motions are needed for the assessment of the degree of stenosis. Moreover, the motion differences between the arteries and atherosclerotic plaque(s) can be associated with plaque instability.

3. ECG region for visualizing ECG waveform: The ECG region is needed for measuring how stenosis and motion patterns of different plaque components change during the cardiac cycle.

Based on the afore-described clinically sensitive regions, the following clinical video quality assessment criteria are used for establishing the reproducibility of the diagnosis:

1. The degree of the artery stenosis: the percentage of the artery that is blocked by the plaque's presence, obscuring blood flow. Significant stenosis can be associated with stroke events.

2. The plaque's morphology [228]: The appearance of the plaque can be used to determine the plaque's type and infer the possible composition of the plaque. The composition of the plaque provides critical information on the risk factors associated with stroke events.

3. The plaque's and artery walls motion characteristics: Plaque motion stability can be classified as concordant or discordant and be used as a potential risk factor as described in [229]–[231]. Here, we note that discordant motion is associated with instability. On the other hand, stiff plaques exhibit concordant motion and tend to be safer. Significant differences between plaque and arterial wall motions can also be used as an indicator of instability.

Individual scores are collected for each of the afore-described clinical criteria. The rating scale allows scores between one (1) at the lowest end, and five (5) on the opposite, highest end. A rating of 5 is assigned to the processed video that is diagnostically equivalent to the original, un-compressed video. A rating of 4 signals the loss of minor clinical details that is still diagnostically acceptable and provides sufficient information for a confident diagnosis. The clinical information present in the processed video is compromised and cannot be trusted for diagnosis purposes when the rating falls below the diagnostically acceptable margin of 4. As a result, a rating of 3, while it still contains clinical information does not qualify for atherosclerosis disease assessment. The lowest clinical rating of 1 corresponds to clinically useless ultrasound video.

4.5 RESULTS AND DISCUSSION

We first present the results of the efficiency of despeckle filtering as a pre-processing step to video coding for ultrasound video communications, followed by video coding standards comparison. In addition to the objective measurements, clinical evaluation is used to verify the clinical capacity of the processed ultrasound videos.

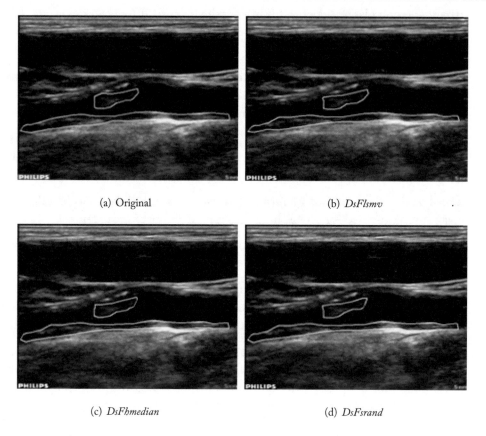

(a) Original   (b) *DsFlsmv*

(c) *DsFhmedian*   (d) *DsFsrand*

Figure 4.3: Original and despeckled ultrasound images examples. Near and far wall atherosclerotic plaque segmentation (outlined by the white lines) using the segmentation algorithm described in [46]. (a) Original, (b) *DsFlsmv*, (c) *DsFhmedian*, (d) *DsFsrad*. Note that the subtle differences between the despeckled and original image are difficult to detect which indicates that the moderate amount of despeckling used here removes higher-frequencies that are not easily detected by the human visual system (HVS) (as desired). They become visible when the clinicians zoom into the regions of interest. Source [212] © *IEEE J. Biomed. Health Inform*, 2011.

4.5.1   CLINICAL ULTRASOUND VIDEO DATASET

The data set is composed of 10 atherosclerotic plaque ultrasound videos, with a spatial resolution of 560×416 at 50 frames per second (fps). Instead of using the QCIF (176×144) and CIF (352×288) resolutions reported in [212], the collected videos at 560×416 do not include any resolution conversions. Video quality assessment is based at this higher resolution exported by the ultrasound equipment. Furthermore, this new set of videos has been specifically collected at

50 fps versus 15 fps to evaluate motion estimation (not covered in [212]). As in [212], to support the reproducibility of the results, we follow an established clinical protocol given in [9].

4.5.2   VIDEO COMPRESSION RESULTS AFTER DESPECKLE FILTERING

The use of despeckle filtering prior to video encoding can lead to significant improvements in rate-distortion performance as demonstrated in Fig. 4.4. As we describe next, these improvements vary significantly depending on the despeckle filtering method.

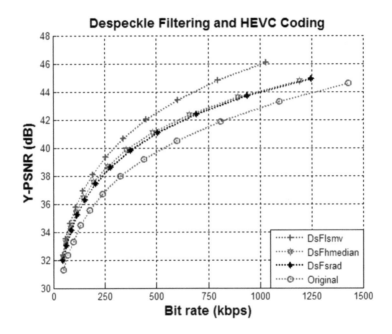

Figure 4.4: Despeckle filtering algorithms efficiency. Rate distortion curves of HEVC encoded videos (mean values of the ten $560 \times 416@50$ fps ultrasound videos for all investigated rate points). All algorithms outperform the conventional encoding procedure involving no speckle filtering. The best performing algorithm is the *DsFlsmv*. The *DsFhmedian* filter marginally outperforms the *DsFsrad*. Source [212] © *IEEE J. Biomed. Health Inform*, 2011.

We present video despeckling examples in Fig. 4.3. Atherosclerotic plaque(s) formed on the near and far wall are outlined using the segmentation algorithm described in [9]. The segmented images allow visualization of the plaque boundaries and plaque morphology. The de-speckled images exhibit very subtle differences that are hard to detect using the human visual system. Note that this is the desired behavior. Ideally, despeckled filtering removes higher frequencies that allow for better compression without visualizing significant artifacts that can compromise the diagnosis. After zooming into the images, it becomes clear that the despeckled images are smoother, missing some of the finer details that are present in the original image. The differences between the different despeckling methods are more difficult to visualize than their differences from the original image.

The differences among the examined methods are easily visualized in the rate-distortion curves of Fig. 4.4. As depicted in the graph, we have significant improvements for all methods. Indicatively, as documented in Table 4.2, *DsFlsmv* reduces bitrate requirements by as much as 43.6% and 39.2%, compared to standard HEVC and H.264/AVC encodings, respectively. The *DsFhmedian* filter lowers bitrate requirements by 34.1% for HEVC and 32.5% for H.264/AVC, while the *DsFsrad* filter achieves bitrate reductions of approximately 23% for both standards. As evident, the trend is the same for both standards. Based on objective evaluation, the *DsFlsmv* is the best performing filter, as it achieves the best PSNR scores while requiring lower bitrates than alternative methods. The hybrid-median marginally outperforms the *DsFsrad* filter. It is important to note here that the objective results regarding the efficiency of the speckle filtering algorithms are also verified by the clinical evaluation as discussed below in Section 4.5.4. To highlight the necessity of efficient video compression methods, we note that for ultrasound video communication purposes, an original uncompressed video would require 93.18 Mbps (560 width resolution x 416 horizontal resolution x 50 frames per second x 8 bits per pixel = 93.18 Mbps). Using the *DsFlsmv* and HEVC encoding for a QP of 28, which achieves diagnostically lossless ultrasound video quality, the transmission rate is reduced to 340 kbps. In other words, a compression ratio of 274 is achieved. Besides the obvious storage space savings, efficient compression that preserves the ultrasound video's clinical capacity allows transmission over existing 3.5G wireless infrastructure (for the particular video resolution), otherwise not feasible for 4G cellular networks (for the uncompressed video).

Table 4.2: Bitrate savings when using despeckle filtering prior to H.264/AVC and HEVC encoding [212]

Despeckle Filtering Method	Bit rate savings relative to	
	H264 Original	HEVC Original
DsFlsmv	39.2%	43.6%
DsFhmedian	32.5%	34.1%
DsFsrad	23.4%	23.5%

4.5.3   VIDEO CODING STANDARDS FOR ULTRASOUND VIDEO COMMUNICATION

More directly, Fig. 4.5 presents a comparison of the median rate-distortion performance of HEVC MP against H.264/AVC HP, H.263 CHC, MPEG-4 ASP, and MPEG-2/H.262 MP. From Fig. 4.5, for the same bitrate, it is clear that HEVC achieves higher levels of video fidelity. As documented in Table 4.3, HEVC achieves average bitrate gains of 33.2% compared to H.264/AVC. The latter is also documented in Fig. 4.6 which shows the dependency of bitrate, PSNR, and SSIM based on the quantization parameter for HEVC MP and H.264/AVC HP. From the plots, for the same quantization parameter value, it is clear that HEVC MP requires less bitrate while achieving higher video quality as measured by PSNR and SSIM. Bitrate savings of 54.6% and 58.3% are observed for earlier H.263 CHC and MPEG-4 ASP standards, respectively, while bitrate gains extend up to 71% when compared to MPEG-2/H.262 MP. Bitrate reductions from using H.264/AVC or other later standards compared to earlier ones are also summarized in Table 4.3.

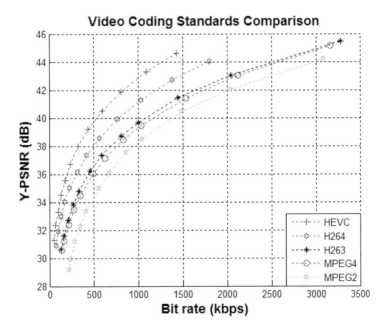

Figure 4.5: Video coding standards comparison. Rate-distortion curves (mean values of the ten 560 × 416@50 fps ultrasound videos for all investigated rate points). HEVC lowers bitrate requirements while it provides for higher PSNR values compared to all prior video coding standards. Source [212] © *IEEE J. Biomed. Health Inform*, 2011.

Table 4.3: Video bitrate savings of different standards compared against previous coding standards

Encoding	Bitrate savings relative to			
	H.264/MPEG-4 AVC HP	H.263 CHC	MPEG-4 ASP	MPEG-2/H.262 MP
HEVC MP	33.2%	54.6%	58.3%	71%
H.264/MPEG-4 AVC HP	-	32.3%	37.7%	56.8%
H.263 CHC	-	-	7.5%	32.4%
MPEG-4 ASP	-	-	-	27.4%

Source [212] © *IEEE J. Biomed. Health Inform*. 2011

4.5.4   CLINICAL EVALUATION

Two medical experts (a cardiovascular surgeon and a neurovascular specialist) were asked to grade the ultrasound videos based on the clinical criteria that were discussed in Section 4.3.6. To emphasize the effects of despeckling, the original videos were presented side-by-side with the despeckled videos. All evaluations were performed using laptop equipment with a spatial resolution 1920×1080 and maximum screen brightness, in a mildly dark environment. Sufficient time was allocated for the medical expert's eyes to adjust to the current lighting conditions. The viewing distance was approximately one meter. Overall, the viewing conditions were comparable to a routine clinical exam.

Table 4.4 summarizes the average scores for the three clinical criteria for the ten ultrasound videos of the data set, prior to compression. As evident in the table, the *DsFlsmv* and *DsFhmedian* filters, yield comparable clinical ratings as the original video. Furthermore, the medical experts emphasized that the overall clinical capacity was neither compromised nor improved from the use of the *DsFlsmv* and *DsFhmedian* filters. On the other hand, in some cases, the *DsFsrad* filter did seem to negatively affect the visualization of the morphology of the plaque as presented in Table 4.4.

Table 4.4: Clinical criteria evaluation for despeckled videos (uncompressed). The average values for stenosis, morphology, and motion are graded from 1 (lowest) to 5 (highest)

Filter	Stenosis	Morphology	Motion
Original video	5.0	5.0	5.0
DFlsmv	4.9	4.7	4.8
DsFhmedian	5.0	4.7	4.9
DsFsrad	5.0	4.2	4.7

Source [212] © *IEEE J. Biomed. Health Inform*. 2011

The clinical capacity of the despeckled and original ultrasound videos following compression was also clinically validated. The results are presented in Table 4.5 for a subset of the examined

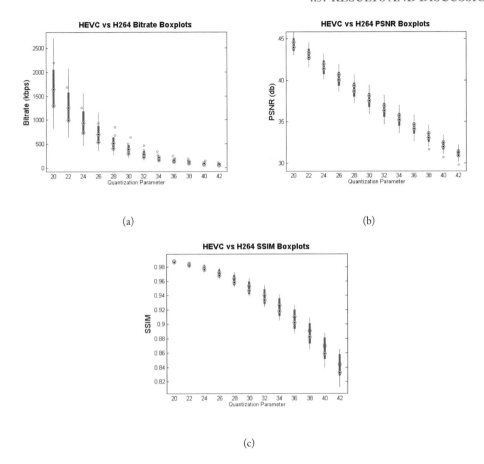

Figure 4.6: HEVC (red) vs. H.264/AVC (blue) comparisons. (a) Bitrate boxplots for different quantization parameters. (b) PSNR scores boxplots. (c) SSIM scores boxplots. (Boxplots use the 10 ultrasound videos of the dataset and all investigated rate points). HEVC requires significantly less bitrate for equivalent quantization levels, while it achieves higher PSNR and SSIM scores than rival H.264/AVC standard. Source [212] © *IEEE J. Biomed. Health Inform*, 2011.

QPs averaged over 4 videos. Clinical scores verify that despeckled ultrasound videos yield comparable diagnostic capacity to the original, non-despeckled videos. Despite documented outliers (mostly for the *DsFsrad* filter), clinical capacity of the compressed videos is enhanced as bitrate budget increases (quantization levels decreases). Overall, ultrasound video denoising can be effectively used to minimize bitrate -and storage- requirements, providing a powerful tool in wireless medical video communications. Artery stenosis and the atherosclerotic plaque's motion assessment gave consistently high ratings. Both criteria relate to the ultrasound video's frame rate and support the requirement of communicating medical videos at the acquired frame rate. Morphology assessment requires encoding using QPs lower than 32 provides for diagnostically acceptable clinical scores (≥ 4). Still, diagnostically acceptable HEVC compression threshold needs to be investigated for a higher number of cases.

Table 4.5: Despeckle filtering ultrasound videos clinical capacity following HEVC encoding. Mean opinion scores are presented for 4 videos for three QPs: 36, 32, and 28

	Stenosis			Morphology			Motion		
Quantization parameter	36	32	28	36	32	28	36	32	28
Original video	4.8	4.8	4.8	4.8	4.8	4.8	4.8	5.0	4.8
DsFlsmv	4.4	4.6	4.6	3.8	4.2	4.6	4.6	4.6	4.6
DsFhmedian	4.6	4.6	4.6	3.8	4.2	4.8	4.4	4.8	4.8
DsFsrad	4.6	4.8	4.4	4.2	4.6	4.4	4.8	4.8	4.8

Source [212] © *IEEE J. Biomed. Health Inform* 2011

4.6 CONCLUDING REMARKS

Both objective evaluations and mean opinion scores based on clinical criteria provide evidence that the emerging HEVC standard yields significant improvements in compression efficiency compared to prior video coding standards. For wired communications channels, we recommend that HEVC be adopted for medical video communication systems. For wireless communications channels, there needs to be an exhaustive evaluation of HEVC's error-resilient performance in noisy channels. Despeckle filtering prior to video encoding can lead to significant bitrate savings without compromising diagnostic quality.

On-going work involves investigating different medical video modalities and emergency trauma videos for (beyond) high-definition encoding using HEVC and transmission over LTE, LTE-Advanced, and WirelessMAN-Advanced wireless networks based on both simulations and real-life scenarios [233]. Future work should also focus on the development of automated methods that can be used to predict the mean-opinion scores on each one of the clinical criteria.

CHAPTER 5

Summary and Future Directions

In this chapter we present the summary findings of the different despeckle filtering algorithms presented in this book by summarising the results presented in Chapters 2 to 4. The summary is based on the ultrasound imaging and video segmentation performance of the IMC and the atherosclerotic carotid plaque (see Chapter 2), on the results for texture analysis and image/video quality evaluation (see Chapter 3) and on the findings for the wireless ultrasound video communication and encoding (see Chapter 4). Furthermore, areas of future research directions on despeckle filtering and briefly discussed.

5.1 SUMMARY FINDINGS ON DESPECKLE FILTERING

Despeckle filtering is an important operation in the enhancement of ultrasound images and videos of the carotid artery, in the case of segmentation, texture analysis, quality evaluation, visual evaluation by the experts and in image and video encoding. In this book a total of 16 despeckle filters were comparatively evaluated on three sets of ultrasound images (see Appendix A.4, 440, 100 and 80) and 10 videos of the common carotid artery and the validation results are summarised in Table 5.1.

As given in Table 5.1, the filter *DsFlsmv*, improved the feature distance (see Table 3.1), the image quality evaluation (see Table 3.4), the optical perception (see Table 3.5A and Table 3.5B) and video transmission for both images and videos of the CCA (see Table 4.2–Table 4.5). The *DsFlsmv* filter also improved the accuracy of the IMC and plaque segmenattion in ultrasound video. The filters *DsFlsminsc*, *DsFhomo*, and *DsFhmedian* improved the feature distance, while the filters *DsFlsmv*, *DsFhmedian* and *DsFkuwahara* also favours video transmission. Filters *DsFlsmv*, *DsFlsminsc*, *DsFgf4d*, *DsFnldif* and *DsFwaveltc* improved the class separation between the asymptomatic and the symptomatic classes (see also Table 3.2) for image despeckling. The filters *DsFgf4d*, *DsFlsminsc* and *DsFhomo* improved the correct classification rate (see also Table 3.3) for image despeckling. Moreover, filters *DsFlsmv*, *DsFgf4d*, *DsFhmedian* and *DsFnldif* gave better image and video quality evaluation results (see Table 3.4, Table 3.6 and Table 3.7). Filters *DsFlsmv*, and *DsFgf4d*, and *DsFhmedian* improved the visual assessment carried out by the experts for images and videos of the CCA (see Table 3.5A, Table 3.5B and Table 3.8). Moreover, the filter *DsFlsmv* improved the accuracy of the IMC and plaque segmentation in ultrasound images and videos (see Chapter 2).

It is clearly shown that filter *DsFlsmv* gave the best performance for both images and videos of the CCA. It is followed by filters *DsFgf4d*, *DsFlsminsc* and *DsFhmedian* (see also Table 5.1).

Table 5.1: Summary findings of despeckle filtering in ultrasound imaging and video of the carotid artery

Despeckle Filter	IMC/Plaque Segmentation: Table 2.1–Table 2.10	Feature distance: Table 3.1 / Table 3.6	Wilcoxon: Table 3.2	kNN classifier: Table 3.3	Image and video quality evaluation: Table 3.4 / Table 3.7	Optical perception evaluation: Table 3.5A, 3.5B / Table 3.8	Video Encoding: Table 4.2–Table 4.5
Image Despeckle Filtering							
Linear Filtering							
DsFlsmv	✓	✓	✓	–	✓	✓	–
DsFlsminsc		✓	✓	✓	–	–	–
Non-Linear Filtering							
DsFgf4d		✓	✓	✓	✓	✓	–
DsFhomo		✓	–	✓	–	–	–
DsFhmedian		✓	–	–	✓	✓	–
Diffusion Filtering							
DsFnldif		–	✓	–	✓	–	–
Wavelet Filtering							
DsFwaveltc		–	✓				–
Video Despeckle Filtering							
DsFlsmv	✓	✓	✓	–	✓	✓	✓
DsFhmedian		✓	✓	–	–	–	✓
DsFkuwahara		–	–	–	–	–	–
DsFsrad		–	–	–	–	–	✓

Filter *DsFlsmv* or *DsFgf4d* can be used for despeckling asymptomatic images where the expert is interested mainly in the plaque composition and texture analysis. Filters *DsFlsmv* or *DsFgf4d* or *DsFlsminsc* of *DsFhmedian* can be used for despeckling of symptomatic images where the expert is interested in identifying the degree of stenosis and the plaque borders. Filters *DsFhomo*, *DsFnldif*, and *DsFwaveltc* gave poorer performance.

Filter *DsFlsmv* gave very good performance, with respect to: (i) preserving the mean and the median as well as decreasing the variance and the speckle index of the image, (ii) increasing the distance of the texture features between the asymptomatic and the symptomatic classes, (iii) significantly changing the SGLDM range of values texture features after filtering based on the Wilcoxon rank sum test, (iv) marginally improving the classification success rate of the kNN classifier for the classification of asymptomatic and symptomatic images in the cases of SF, SMF and TEM feature sets, (v) improving the image quality, and (vi) video encoding where it may lead to significant bitrate savings without compromising diagnostic quality. The *DsFlsmv* filter, which is a simple filter, is based on local image statistics. It was first introduced in [6, 19, 27] by Jong-Sen Lee and co-workers and it was tested on a few SAR images with satisfactory results. It was also used for SAR imaging in [18] and image restoration in [37], again with satisfactory results. More recently, the filter *DsFlsmv* was introduced in [2] where it was integrated in a despeckle image (IDF) despeckle filtering software toolbox together with 16 different despeckle filtering methods. Moreover, the despeckle filter *DsFlsmv* was also recently introduced in a despeckle video filtering software toolbox (VDF) [3] together with three other despeckle filters (see also Table 5.1).

Filter *DsFlsminsc* gave the best performance with respect to: (i) preserving the mean, as well as decreasing the variance and the speckle index and increasing the contrast of the image and video, (ii) increasing the distance of the texture features between the asymptomatic and the symptomatic classes, (iii) significantly changing the SGLDM texture features after filtering based on the Wilcoxon rank sum test, (iv) improving the classification success rate of the kNN classifier for the classification of asymptomatic and symptomatic images in the cases of SF, SGLDMr, GLDS, NGTDM, FDTA and FPS feature sets. Filter *DsFlsminsc* was originally introduced by Nagao in [38] and was tested on an artificial and a SAR image with satisfactory performance. In this study the filter was modified, by using the speckle index instead of the variance value for each sub window (as described in Volume I of this book [1], Section 3.3 and Equation 2.13).

Filter *DsFgf4d* gave very good performance with respect to: (i) decreasing the speckle index, (ii) marginally increasing the distance of the texture features between the asymptomatic and the symptomatic classes, (iii) significantly changing the SGLDM range of values texture features after filtering based on the Wilcoxon rank sum test, and (iv) improving the classification success rate of the kNN classifier for the classification of asymptomatic and symptomatic images in the cases of SGLDMm, GLDS, NGTDM, SFM and TEM feature sets. The geometric filter *DsFgf4d* was introduced by Crimmins [15], and was tested visually on a few SAR images with satisfactory results.

The results of our study showed, that observer variability, and sensitivity are important in image quality evaluation, and can only be compensated when assessments are made against a standard scale of quality, such as the image quality evaluation metrics proposed in this study. Observer variability may also be compensated by additional tests employing image quality and texture measures, as proposed in this study, for quantifying image quality.

A total of 16 different despeckle filters were documented in this book based on linear filtering, non-linear filtering, diffusion filtering and wavelet filtering. We have evaluated despeckle filtering on 440 (220 asymptomatic and 220 symptomatic) ultrasound images of the carotid artery bifurcation, based on visual evaluation by two medical experts, texture analysis measures, and image quality evaluation metrics. A linear despeckle filter based on local statistics (*DsFlsmv*) improved the class separation between the asymptomatic and the symptomatic classes, gave only a marginal improvement in the percentage of correct classifications success rate based on texture analysis and the kNN classifier, and improved the visual assessment by the experts. It was also found that the *DsFlsmv* despeckle filter can be used for despeckling asymptomatic images where the expert is interested mainly in the plaque composition and texture analysis, whereas a geometric despeckle filter (*DsFgf4d*) can be used for despeckling of symptomatic images where the expert is interested in identifying the degree of stenosis and the plaque borders. The results of this study, suggest that the first order statistics despeckle filter *DsFlsmv*, may be applied on ultrasound images to improve the visual perception and automatic image analysis.

Furthermore, despeckle filtering was investigated as a pre-processing step for the automated segmentation of the IMT [98] and the carotid plaque [9], followed by the carotid plaque texture analysis, and classification (as documented in the above paragraph). Despeckle filters *DsFlsmv*, *DsFlsminsc*, and *DsFgf4d* gave the best performance for the segmentation tasks. It was shown in [98] that when normalization and speckle reduction filtering is applied on ultrasound images of the carotid artery prior to IMT segmentation, the automated segmentation measurements are closer to the manual measurements. This field has also been investigated by our group [147]. Our findings showed promising results, however, further work is required to evaluate the performance of the suggested despeckle filters at a larger scale as well as their impact in clinical practice. In addition, the usefulness of the proposed despeckle filters, in portable ultrasound systems and in wireless telemedicine systems still has to be investigated.

Our results on image quality evaluation (for comparing two different ultrasound scanners, ATL HDI-3000 and ATL HDI-5000), showed that normalization and speckle reduction filtering are important pre-processing steps favouring image quality. Additionally, the usefulness of the proposed methodology based on quality evaluation metrics combined with visual evaluation in ultrasound imaging and in wireless telemedicine systems needs to be further investigated.

The results presented in this book on video despeckling showed that normalization and despeckle filtering improves the outcome of the IMT segmentation and produces more accurate and reproducible results when compared with the manual segmentation method. Video despeckle filtering also increases the optical perception evaluation by experts and may also also lead to significant bitrate savings without compromising diagnostic quality prior to video encoding.

> **Key Message**: For those readers whose principal need is to use existing image despeckle filtering technologies and apply them on different type of images or video, there is no simple answer regarding which specific filtering algorithm should be selected without a significant understanding of both the filtering fundamentals, and the application environment under investigation. A number of issues would need to be addressed. These include availability of the image/video to be processed/analyzed, the required level of filtering, the application scope (general-purpose or application-specific), the application goal (for extracting features from the image or for visual enhancement), the allowable computational complexity, the allowable implementation complexity, and the computational requirements (e.g., real-time or offline). We believe that a good understanding of the contents of this book can help the readers make the right choice of selecting the most appropriate filter for the application under development. Furthermore, the despeckle filtering evaluation protocol documented in Table 1.4 could also be exploited.

5.2   FUTURE DIRECTIONS

As it has already been documented in the corresponding section on future directions in the companion volume of the book [1], the despeckle filtering algorithms, and the measures for image quality evaluation introduced in this book can also be generalised and applied to other image and video processing applications. Only a small number of filtering algorithms and image quality evaluation metrics were investigated in this book, and numerous extensions and improvements can be envisaged. Most importantly more comparative studies of despeckle filtering are necessary, where different filters could be evaluated by multiple experts as well as based on image quality and evaluation metrics as also proposed in this book.

Additionally, the issue of video despeckling is still in its infancy and this area of research work will receive significant attention in the following years. Although it is noted that the proposed methodology and filtering algorithms documented in this book may be also investigated in video sequences (by frame filtering). There are many issues related to video despeckle filtering that remain to be solved. In general, the development of a multiplicative model based on video sequences is required, since most of the models developed for video filtering were for additive noise [182–185].

Despeckle filtering may be also applied in the pre-processing of ultrasound images for other organs, including the detection of hyperechoic or hypoechoic lesions in the kidney, liver, spleen, thyroid, kidney, echocardiographic images, mammography and other. It may be particularly effective when combined with harmonic imaging, since both can increase tissue contrast. Speckle reduction can also be extremely valuable when attempting to fuse ultrasound with Computed Tomography (CT), MRI, Positron Emission Tomography (PET) or Optical Coherence Tomography (OCT) images. For example, when a lesion is suspected on a CT scan but it is not clearly visible ultrasound despeckle filtering can be applied in order to accentuate subtle borders that may be masked by speckle.

Great efforts are also currently made in optimizing the despeckle filtering algorithms in order to achieve better performance [80, 234]. It is foreseen that optimization of a despeckling algorithm would be dependent on transducer geometry, operating frequency, focal point(s), distribution of pixel values due to speckle, subject being scanned, etc. Automatic selection of optimal despeckling algorithm's parameters would provide a useful tool for research and clinical applications.

Ultrasound imaging instrumentation, linked with imaging hardware and software technology have been rapidly advancing in the last two decades. Although these advanced imaging devices produce higher quality images and video, the need still exists for better image and video processing techniques including despeckle filtering. Towards this direction, it is anticipated that the effective use of despeckle filtering (by exploiting the filters and algorithms documented in this book) will greatly help in producing images with higher quality. These images that would be not only easier to visualise and to extract useful information, but would also enable the development of more robust image pre-processing and segmentation algorithms, minimizing routine manual image analysis and facilitating more accurate automated measurements of both industrially and clinically-relevant parameters.

APPENDIX A

Appendices

Appendix Section A.1 contains a listing of all the functions included in the image despeckle filtering (IDF) toolbox, as introduced in Table 1.3 and Volume I of this book. The IDF toolbox, which can be downloaded from `http://www.medinfo.cs.ucy.ac.cy/`, includes all the functions used for the texture analysis, (see also companion monograph), as well as for the image quality and evaluation. Appendix A.2 lists all the functions included in the video despeckle filtering (VDF) toolbox for video analysis introduced in Section 2.2. Appendix A.3 presents an example in Matlab™ code for a complete application of despeckling, image quality evaluation and texture analysis. Appendix A.4 presents an example in Matlab™ code for a complete application of despeckling, video quality evaluation and All page numbers listed refer to pages in the book, indicating where a function is first used and illustrated.

A.1 DESPECKLE FILTERING, TEXTURE ANALYSIS, AND IMAGE QUALITY EVALUATION TOOLBOX FUNCTIONS (IDF TOOLBOX)

The following MATLAB functions are grouped in categories as presented in Table 1.2 of this book and introduced in the companion Vol. I of this book [1].

Function Category and Name	Description	Page or Other Location in [1]
Linear Filtering		
DsFlsmv	Mean and variance local statistics despeckle filter	p. 43, Algorithm 3.1
DsFlsmvsk1d	Minimum variance homogenous 1D mask despeckle filter	p. 56, [1]
DsFlsmvsk2d	Mean variance, higher moments local statistics despeckle filter	p. 58, Algorithm 3.3
DsFlsminsc	Minimum speckle index homogenous mask despeckle filter	p. 61, Algorithm 3.4
DsFwiener	Wiener despeckle filter	p. 56, Algorithm 3.2
Non-Linear Filtering		

DsFmedian	Median despeckle filter	p. 68, Algorithm 4.1
DsFls	Linear scaling of the gray level values despeckle filter	p. 68
DsFca	Linear scaling of the gray-levels despeckle filter	p. 69, Algorithm 4.2
DsFlecasort	Linear scaling and sorting despeckle filter	p. 71
DsFgf4d	Geometric despeckle filtering	p. 74, Algorithm 4.4
DsFhomog	Most homogeneous neighbourhood despeckle filter	p. 71, Algorithm 4.3
DsFhomo	Homomorphic despeckle filtering	p. 74
DsFhmedian	Hybrid median despeckle filter	p. 80, Algorithm 4.6
DsFkuwahara	Kuwahara nonlinear despeckle filtering	p. 82, Algorithm 4.7
DsFnlocal	Non-local despeckle filtering	p. 84, Algorithm 4.8
Diffusion Filtering		
DsFad	Perona and Malik diffusion filter	p. 85
DsFsrad	Speckle reducing anisotropic diffusion filter	p. 87, Algorithm 5.1
DsFnldif	Nonlinear coherent diffusion despeckle filter	p. 93
DsFncdif	Nonlinear complex diffusion filtering	p. 95
Wavelet Filtering		
DsFwaveltc	Wavelet despeckle filtering	p. 99, Algorithm 6.1

The following image texture analysis MATLAB functions (also presented in the companion monograph), which are included in the IDF [2] and VDF [3] toolboxes, for image and video analysis are here below described:

Function Category and Name	Description	Page or Other Location [1]
DsTnwfos	First-Order Statistics (FOS) (features 1–5)	Website
	Texture Analysis Functions	
DsTnwsgldm	Haralick Spatial Gray Level Dependence Matrices (SGLDM) (6–31)	Website
DsTnwgldmc	Gray Level Difference Statistics (GLDS) (32–35)	Website
DsTnwngtdmn	Neighbourhood Gray Tone Difference Matrix (NGTDM) (36–40)	Website
DsTnwsfm	Statistical Feature Matrix (SFM) (41–44)	Website
DsTnlaws	Laws Texture Energy Measures (TEM) (45–50)	Website
DsTfdta2	Fractal Dimension Texture Analysis (FDTA) (51–54)	Website

DsTfps	Fourier Power Spectrum (FPS) (55–56)	Website
DsTfshape2	Shape (x, y, area, perim, perim^2/area) (57–61)	Website
DsTintens2	Intensity difference vector with step s	Website
DsTleast	Estimation of the curve slope using least squares	Website
DsTresol2	Multiple resolution feature extraction	Website
DsTexfeat	Main texture analysis function	Website

Website: The Matlab™ code can be downloaded from `http://www.medinfo.cs.ucy.ac.cy` as well as from Researchgate.

The following image quality evaluation MATLAB functions are given as presented in the companion monograph to this book and are included in the IDF [2] and VDF [3] toolboxes for image and video analysis.

Function Category and Name	Description	Page or Other Location in [1]
Quality Evaluation		
DsQEgae	Geometric average error	p. 35
DsQEmse	Mean square error	p. 38
DsQEsnr	Signal-to-noise ratio	p. 36
DsQErmse	Randomized mean square error	p. 35
DsQEpsnr	Peak signal-to-noise radio	p. 36
DsQEminkowski	Minkowski metrics, 3rd (M3) and 4th (M4) moments	p. 35
DsQEimg_qi	Universal quality index	p. 36
DsQEssim_index	Structural similarity index	p. 36
DsQEget_dir_files	Get directory files	Website
DsQE_quality_evaluation	Main quality evaluation program	Website
DsQmetrics	Function for running all above Quality Evaluation Metrics	Website

Website: The Matlab™ code can be downloaded from `http://www.medinfo.cs.ucy.ac.cy`

A.2 DESPECKLE FILTERING, TEXTURE ANALYSIS, AND VIDEO (VDF TOOLBOX) QUALITY EVALUATION TOOLBOX FUNCTIONS

The following Matlab™ functions are included in the VDF toolbox [3] used for video despeckle filtering.

Function Category and Name	Description	Page or Other Location in [1]
Linear filtering		
DsFlsmv	Mean and variance local statistics despeckle filter	p. 43, Algorithm 3.1
Nonlinear Filtering		
DsFhmedian	Hybrid median despeckle filter	p. 80, Algorithm 4.6
DsFkuwahara	Kuwahara nonlinear despeckle filtering	p. 82, Algorithm 4.7
Diffusion Filtering		
DsFsrad	Speckle reducing anisotropic diffusion filter	p. 88, Algorithm 5.1
Wavelet Filtering		
DsFwaveltc	Wavelet despeckle filtering	p. 99, Algorithm 6.1

A.3   EXAMPLES OF RUNNING THE DESPECKLE FILTERING TOOLBOX FUNCTIONS

Matlab™ code for the **DsQmetrics.m** function:

The following code sequence will read an image and apply the *DsFlsmv* despeckle filter on the image iteratively five times, by using a moving sliding window of 7×7 pixels. The texture features as well as the image quality metrics between the original and the despeckled images, are calculated with the code in A.1 and A.2 (included in the IDF toolbox [2]), and stored in the variable matrix A and B, respectively. The image quality metrics between the original and the despeckled images are stored in the matrix M.

A.4   MATERIAL AND RECORDING OF ULTRASOUND IMAGES AND VIDEOS

Three different imaging datasets and one video dataset were used for our experiments carried out in this book and are here below described. The first imaging dataset was used for evaluating despeckle filtering, the second for evaluating the image quality of two ultrasound scanners. The third image database was acquired from normal subjects and was used for comparison purposes. The video dataset was used to evaluate despeckle filtering on videos.

Dataset 1: The images of the carotid artery bifurcation used for the despeckling (the first image dataset) were acquired using the ATL HDI-3000 ultrasound scanner. The ATL HDI-3000 ultrasound scanner is equipped with 64 elements fine pitch high-resolution, 38 mm broadband array, a multi element ultrasound scan head with an operating frequency range of 4–7 MHz, an acoustic aperture of 10×8 mm and a transmission focal range of 0.8–11 cm [185]. All images were recorded as they are displayed in the ultrasound monitor, after logarithmic compression. The images were recorded digitally on a magneto optical drive, with a resolution of 768×756 pixels with 256 gray levels. The image resolution was 16.66 pixels/mm. B-mode scan settings were adjusted so that the maximum dynamic range was used with a linear post-processing curve. The position of the probe was adjusted so that the ultrasonic beam was vertical to the artery wall. The time gain compensation, TGC, curve was adjusted, (gently sloping), to produce uniform intensity of echoes on the screen, but it was vertical in the lumen of the artery where attenuation in blood was minimal, so that echogenicity of the far wall was the same as that of the near wall. The overall gain was set so that, the appearance of the plaque was assessed to be optimal, and slight noise appeared within the lumen. It was then decreased so that at least some areas in the lumen appeared to be free of noise (black).

The first image dataset used for despeckle filtering, consisted of a total of 440 ultrasound images of the carotid artery bifurcation, 220 asymptomatic and 220 symptomatic and were acquired with the ATL HDI-3000 scanner. Asymptomatic images were recorded from patients at risk of atherosclerosis in the absence of clinical symptoms, whereas symptomatic images were recorded from patients at risk of atherosclerosis, which have already developed clinical symptoms, such as a stroke episode.

Dataset 2: The second image dataset consisted of a total of 80 symptomatic B-mode longitudinal ultrasound images, used for the image quality evaluation and plaque segmentation, from identical vessel segments of the carotid artery bifurcation, were acquired from each ultrasound scanner (from the ATL HDI-3000 and the ATL HDI-5000 scanner). The images were recorded digitally on a magneto optical drive with a resolution of 768×576 pixels with 256 gray levels. These images were recorded at the Institute of Neurology and Genetics, in Nicosia, Cyprus, from 32 female and 48 male symptomatic patients aged between 26 and 95 years old, with a mean age of 54 years old. These subjects were at risk of atherosclerosis, which have already developed clinical symptoms, such as a stroke or a transient ischemic attack. In addition, 10 symptomatic ultrasound images of the carotid artery representing different types of atherosclerotic carotid plaque formation with irregular geometry typically found in this blood vessel were acquired from each scanner. The images used for the image quality evaluation (in the second image database) were acquired with the ATL HDI-5000 ultrasound scanner, which is equipped with a 256 elements fine pitch high-resolution 50 mm linear array, a multi element ultrasound scan head with an extended operating frequency range of 5–12 MHz and it offers real spatial compound imaging. The scanner increases the image clarity using SonoCT imaging by enhancing the resolution and borders, and interface margins are better displayed. Several tests made by the manufacturer showed that, the ATL HDI-5000 scanner was overall superior to conventional two-dimensional imaging systems, primarily because of the reduction of speckle, contrast resolution, tissue differentiation, and the image was visually better [185].

Digital images were resolution normalized at 16.66 pixels/mm [1, 46, 48]. The images were recorded at the Saint Mary's Hospital, Imperial College of Medicine, Science and Technology, UK, from 32 female and 48 male symptomatic patients aged between 26 and 95 years old, with a mean age of 54 years. These subjects were at risk of atherosclerosis and have already developed clinical symptoms, such as a stroke or a transient ischemic attack (TIA).

Dataset 3: The third image dataset included 100 ultrasound B-mode ultrasound images of the CCA (44 women and 56 men) acquired from normal subjects with no atherosclerotic disease at the Institute of Neurology and Genetics, in Nicosia, Cyprus [2, 186]. They had a mean $\pm$SD age of 38.11 ± 7.56 years. These were acquired using the ATL HDI-3000 ultrasound scanner (Advanced Technology Laboratories, Seattle, USA), with a linear probe (L74), with a recording frequency of 7 MHz, a velocity of 1550 m/s and 1 cycle per pulse, which resulted to a wavelength (spatial pulse length) of 0.22 mm and an axial system resolution of 0.11 mm. They were recorded digitally on a magneto optical drive with a resolution of 768×576 pixels with 256 gray levels. The technical characteristics of the ultrasound scanner (multi element ultrasound scan head, operating frequency, acoustic aperture, and transmission focal range) have already been published in [48, 100, 102]. The images were logarithmically compressed and were recorded digitally on a magneto optical drive at size of 768×576 pixels with 256 gray levels. The images were recorded at the Cyprus Institute of Neurology and Genetics, in Nicosia, Cyprus, from 42 female and 58 male asymptomatic subjects aged between 26 and 95 years old, with a mean age of 54 years.

Dataset 4: The video dataset consisted of a total of 43 B-mode longitudinal ultrasound videos (from 38 asymptomatic subjects, aged 56 ± 12 and 5 symptomatic, aged 53 ± 16 subjects, 17 female and 26 male) of the CCA bifurcation. The videos were recorded from subjects representing different types of atherosclerotic plaque formation with irregular geometry typically found in this blood vessel. Almost all subjects demonstrated a left or a right CCA stenosis of 30% or larger. The videos were acquired by the ATL HDI-5000 ultrasound scanner (Advanced Technology Laboratories, Seattle, USA) and were recorded digitally on a magneto optical drive, with a size of 576×768 pixels with 256 gray levels, a pixel size of 59 μm (17 pixels per mm) and having a frame rate of 100 frames per second. This frame rate is high, however it was used as these videos will also be analysed for plaque motion estimation. The ATL HDI-5000 ultrasound scanner is equipped with a 256-element fine pitch high-resolution 50 mm linear array, a multi-element ultrasound scan head with an extended operating frequency range of 5–12 MHz and it offers real spatial compound imaging. The video segmentations were performed for 3–5 seconds intervals, covering in general 2–3 cardiac cycles.

Code A.1

```
function f=DsQmetrics(I,K);
% I: Original input noisy image
% K: Despeckled input Image
% f: Matrix with image quality metrics

I=double(I); K=double(K);
 gaer = DsQgae(I,K);
 metrics= [gaer];

% calculate the mean square error mse
mser=DsQmse(I,K);
metrics=[metrics, mser];

% calculate the signal-to-noise radio snr
snrad=DsQsnr(I,K);
metrics=[metrics, snrad];

% calculate the square root of the mean square error
rmser=DsQrmse(I, K);
metrics=[metrics, rmser];

% Calculate the peak-signal-to-noise radio
psnrad=DsQpsnr(I,K);
metrics=[metrics, psnrad];

% Calculate the Minkwofski measure
[M3, M4] = DsQminkowski(I, K);
metrics=[metrics, M3, M4];

% Calculate the universal quality index
[quality, quality_map] = DsQimg_qi(I,K);
metrics=[metrics, quality];

% Calculate the structural similarity index
[mssim, ssim_map] = DsQssim_index(I, K);
metrics=[metrics, mssim];

% calculate aditional metrics
[MSE,PSNR,AD,SC,NK,MD,LMSE,NAE,PQS]= DsQiq_measures(I,K);
metrics=[metrics, AD, SC, NK, MD, LMSE, NAE, PQS];
 f= metrics;
```

Code A.2

```
% Read the image original.tif and store it in variable image
image = imread ('original.tif');

% Apply the despeckle filter DsFlsmv on the image using a sliding moving window of 7x7 pixel,
%iteratively 5 times
despeckle = DsFlsmv (image, [7 7], 5);
% Show the original and the despeckled images on the screen
figure, imshow (image); figure, imshow (despeckle);

% Calculate the texture features for the original and the despeckled images
Orig_textfeat = DsTexfeat (image);
Desp_textfeat = DsTexFeat (despeckle);

% Save the extracted features of the original and despeckled images in the mat files A and B
save Orig_textfeat A;
save Desp_textfeat B;

% Calculate the image quality evaluation metrics between the original and the despeckled images
% and save them in a matrix M
M = DsQmetrics (image, despeckle);

%The mat files A, B can then be loaded into the MATLAB workspace for opening, reading and
storing the features and image quality metrics. This can be made by double clicking the mat files.
% The command whos will show the files loaded
whos
% The open command will then open the file A, B and M
open A;
open B;
open M;
%The texture features for both the original and despeckled images and the quality evaluation
%metrics can now be manipulated or saved elsewhere.
```

References

[1] C.P. Loizou and C.S. Pattichis, "Despeckle filtering for ultrasound imaging and video, Volume I: Algorithms and software," 2nd ed., Synthesis lectures on algorithms and software in engineering, Morgan & Claypool Publishers, 1537 Fourth street, Suite 228, San Rafael, CA 94901, vol. 7, no. 1, pp. 1–180, April 2015. DOI: 10.2200/S00116ED1V01Y200805ASE001. 1, 2, 3, 4, 11, 18, 20, 32, 34, 35, 55, 58, 63, 66, 70, 71, 73, 74, 85, 107, 119, 121, 123, 128

[2] C.P. Loizou, C. Theofanous, M. Pantziaris, and T. Kasparis, "Despeckle filtering software toolbox for ultrasound imaging of the common carotid artery," *Comput. Meth. & Progr. Biomed.*, vol. 114, no. 1, pp. 109–124, 2014. DOI: 10.1016/j.cmpb.2014.01.018. 1, 9, 11, 13, 18, 32, 33, 55, 56, 70, 71, 86, 119, 124, 125, 126, 128

[3] C.P. Loizou, C. Theofanous, M. Pantziaris, T. Kasparis, P. Christodoulides, A.N. Nicolaides, and C.S. Pattichis, "Despeckle filtering toolbox for medical ultrasound video," *Int. J. Monitoring & Surveill. Technol. Resear. (IJMSTR): Special issue Biomed. Monitor. Technol.*, vol. 4, no. 1, pp. 61–79, 2013. DOI: 10.4018/ijmstr.2013100106. 1, 9, 11, 14, 20, 55, 56, 70, 71, 73, 85, 86, 119, 124, 125

[4] Z. Wang, A. Bovik, H. Sheikh, and E. Simoncelli, "Image quality assessment: From error measurement to structural similarity," *IEEE Trans. Image Proces.*, vol. 13, no. 4, pp. 600–612, 2004. DOI: 10.1109/TIP.2003.819861. 1, 73, 80, 81

[5] T. Elatrozy, A. Nicolaides, T. Tegos, A. Zarka, M. Griffin, and M. Sabetai, "The effect of B-mode ultrasonic image standardization of the echodensity of symptomatic and asymptomatic carotid bifurcation plaque," *Int. Angiology*, vol. 17: 179–186, no. 3, 1998. 1, 17, 18, 19, 79, 86, 93

[6] J.S. Lee, "Speckle analysis and smoothing of synthetic aperture radar images," *Computer Graph. & Image Proces.*, vol. 17, pp. 24–32, 1981. DOI: 10.1016/S0146-664X(81)80005-6. 1, 73, 74, 119

[7] C.P. Loizou, C.S. Pattichis, C.I. Christodoulou, R.S.H. Istepanian, M. Pantziaris, and A. Nicolaides "Comparative evaluation of despeckle filtering in ultrasound imaging of the carotid artery," *IEEE Trans. Ultr. Ferroel. & Freq. Contr.*, vol. 52, no. 10, pp. 1653–1669, 2005. DOI: 10.1109/TUFFC.2005.1561621. 9, 11, 33, 35, 51, 53, 70, 71, 72, 73, 74, 79, 81, 85, 86

[8] C.P. Loizou, C.S. Pattichis, M. Pantziaris, T. Tyllis, and A. Nicolaides, "Quantitative quality evaluation of ultrasound imaging in the carotid artery," *Med. Biol. Eng. Comput.*," vol. 44, no. 5, pp. 414–426, 2006. DOI: 10.1007/s11517-006-0045-1. 1, 9, 11, 53, 72, 73, 75, 76, 77, 78, 79, 81, 85, 88, 89

[9] C.P. Loizou, C.S. Pattichis, M. Pantziaris, and A. Nicolaides, "An integrated system for the segmentation of atherosclerotic carotid plaque," *IEEE Trans. Inform. Techn. Biomed.*," vol. 11, no. 5, pp. 661–667, 2007. DOI: 10.1109/TITB.2006.890019. 11, 31, 32, 33, 34, 35, 36, 37, 38, 39, 40, 50, 53, 73, 79, 80, 111, 120

[10] T. Greiner, C.P. Loizou, M. Pandit, J. Mauruschat, and F.W. Albert, "Speckle reduction in ultrasonic imaging for medical applications," *Proc. ICASSP91, Int. Conf. Acoustic Signal speech and Processing*, Toronto Canada, May 14–17, pp. 2993-2996, 1991. DOI: 10.1109/ICASSP.1991.151032.

[11] C.P. Loizou and C.S. Pattichis, "Despeckle filtering algorithms and Software for Ultrasound Imaging," Synthesis Lectures on Algorithms and Software for Engineering, Morgan & Claypool Publishers, San Rafael, CA, 2008. DOI: 10.2200/S00116ED1V01Y200805ASE001. 9, 17, 18, 19, 32, 33, 34, 35, 67, 68, 70, 71, 86

[12] C.P. Loizou, V. Murray, M.S. Pattichis, M. Pantziaris, A.N. Nicolaides, and C.S. Pattichis, "Despeckle filtering for multiscale amplitude-modulation frequency-modulation (AM-FM) texture analysis of ultrasound images of the intima-media complex," *Int. J. Biomed. Imag.*, vol. 2014, Art. ID. 518414, 13 pages, 2014. DOI: 10.1155/2014/518414.

[13] C.P. Loizou, S. Petroudi, C.S. Pattichis, M. Pantziaris, and A.N. Nicolaides, "An integrated system for the segmentation of atherosclerotic carotid plaque in ultrasound video," *IEEE Trans. Ultras. Ferroel. Freq. Contr.*, vol. 61, no. 1, pp. 86–101, 2014. DOI: 10.1109/TUFFC.2014.6689778. 20, 34, 35, 36, 39, 40, 45, 46, 47, 48, 86

[14] C.P. Loizou, T. Kasparis, P. Christodoulides, C. Theofanous, M. Pantziaris, E. Kyriakou, and C.S. Pattichis, "Despeckle filtering in ultrasound video of the common carotid artery," *12th Int. Conf. Bioinf. & Bioeng. Proc. (BIBE)*, Larnaca, Cyprus, Nov. 11–13, pages 4, 2012. DOI: 10.1109/BIBE.2012.6399756. 1, 11, 20, 41, 42, 43, 44, 70, 86

[15] L. Busse, T.R. Crimmins, and J.R. Fienup, "A model based approach to improve the performance of the geometric filtering speckle reduction algorithm," *IEEE Ultrasonic Symposium*, pp. 1353–1356, 1995. DOI: 10.1109/ULTSYM.1995.495807. 1, 119

[16] D.T. Kuan, A.A. Sawchuk, T.C. Strand, and P. Chavel, "Adaptive restoration of images with speckle," *IEEE Trans. Acoustic Speech & Signal Processing*, vol. ASSP-35, pp. 373–383, 1987. DOI: 10.1109/TASSP.1987.1165131.

[17] M. Insana et al., "Progress in quantitative ultrasonic imaging," *SPIE Vol. 1090 Medical Imaging III, Image Formation*, pp. 2–9, 1989. DOI: 10.1117/12.953184. 73

[18] V.S. Frost, J.A. Stiles, K.S. Shanmungan, and J.C. Holtzman, "A model for radar images and its application for adaptive digital filtering of multiplicative noise," *IEEE Trans. Pattern Analysis & Machine Intelligence*, vol. 4, no. 2, pp.157–165, 1982. DOI: 10.1109/TPAMI.1982.4767223. 70, 119

[19] J.S. Lee, "Digital image enhancement and noise filtering by using local statistics," *IEEE Trans. Pattern Analysis & Machine Intelligence*, PAMI-2, no. 2, pp. 165–168, 1980. DOI: 10.1109/TPAMI.1980.4766994. 33, 70, 71, 72, 73, 79, 119

[20] D. Lamont, L. Parker, M. White, N. Unwin, et al., "Risk of cardiovascular disease measured by carotid intima-media thickness at age 49–51: life course study," *BMJ*, vol. 320, pp. 273-278, 29 Jan. 2000. DOI: 10.1136/bmj.320.7230.273.

[21] C.B. Burckhardt, "Speckle in ultrasound B-mode scans," *IEEE Trans. Sonics & Ultrasonics*, vol. SU-25, no. 1, pp. 1–6, 1978. DOI: 10.1109/T-SU.1978.30978.

[22] R.F. Wagner, S.W. Smith, J.M. Sandrik, and H. Lopez, "Statistics of speckle in ultrasound B-scans," *IEEE Trans. Sonics Ultrasonics*, vol. 30, pp. 156–163, 1983. DOI: 10.1109/T-SU.1983.31404. 74

[23] J.W. Goodman, "Some fundamental properties of speckle," *J. Optical Society America*, vol. 66, no. 11, pp. 1145–1149, 1976. DOI: 10.1364/JOSA.66.001145.

[24] Y. Yongjian and S.T. Acton, "Speckle reducing anisotropic diffusion," *IEEE Trans. Image Processing*, vol. 11, no. 11, pp. 1260–1270, 2002. DOI: 10.1109/TIP.2002.804276. 70, 71, 72, 73

[25] R.W. Prager, A.H. Gee, G.M. Treece, and L. Berman, "Speckle detection in ultrasound images using first order statistics," University of Cambridge, Dept. of Engineering, GUED/F-INFENG/TR 415, pp. 1–17, 2002.

[26] C.I. Christodoulou, C. Loizou, C.S. Pattichis, M. Pantziaris, E. Kyriakou, M.S. Pattichis, C.N. Schizas, and A. Nicolaides, "Despeckle filtering in ultrasound imaging of the carotid artery," *2nd Joint EMBS/BMES Conference*, Houston, TX, pp. 1027–1028, Oct. 23–26, 2002. DOI: 10.1016/j.cmpb.2014.01.018. 11, 70, 74

[27] J.S. Lee, "Refined filtering of image noise using local statistics," *Computer Graphics and Image Processing*, vol. 15, pp. 380–389, 1981. DOI: 10.1016/S0146-664X(81)80018-4. 1, 35, 119

[28] V. Dutt, "Statistical analysis of ultrasound echo envelope," Ph.D. dissertation, Mayo Graduate School, Rochester, MN, 1995. 1, 70

[29] M. Amadasun and R. King, "Textural features corresponding to textural properties," *IEEE Trans. Systems, Man, & Cybernetics*, vol. 19, no. 5, pp. 1264–1274, 1989. DOI: 10.1109/21.44046. 56

[30] C.M. Wu, Y.C. Chen, and K.-S. Hsieh, "Texture features for classification of ultrasonic images," *IEEE Trans. Med. Imaging*, vol. 11, pp. 141–152, 1992. DOI: 10.1109/42.141636. 56, 57, 61, 63, 70, 80

[31] T.J. Chen, K.S. Chuang, Jay Wu, S.C. Chen, I.M. Hwang, and M.L. Jan, "A novel image quality index using Moran I statistics," *Physics in Medicine and Biology*, vol. 48, pp. 131–137, 2003. DOI: 10.1088/0031-9155/48/8/402. 74

[32] S. Winkler, "Vision models and quality metrics for image processing applications," Ph.D., University of Lausanne-Switzerland, 2000. 1, 80, 81

[33] C. Christodoulou, C. Pattichis, M. Pantziaris, and A. Nicolaides, "Texture-Based Classification of Atherosclerotic Carotid Plaques," *IEEE Trans. Medical Imaging*, vol. 22, no. 7, pp. 902–912, 2003. DOI: 10.1109/TMI.2003.815066. 1, 56, 57, 61, 70, 72, 80

[34] E. Kyriakou, M.S. Pattichis, C.I. Christodoulou, C.S. Pattichis, S. Kakkos, M. Griffin, and A.N. Nicolaides, "Ultrasound imaging in the analysis of carotid plaque morphology for the assessment of stroke," in *Plaque Imaging: Pixel to molecular level*, Eds. Suri, J.S., Yuan, C., Wilson, D.L., and Laxminarayan, S., IOS press, pp. 241–275, 2005. 1, 53, 80, 86

[35] J. Grosby, B.H. Amundsen, T. Hergum, E.W. Remme, S. Langland, and H. Trop, "3D Speckle tracking for assessement of regional left ventricular function," *Ultrasound Med. Biol.*, vol. 35, pp. 458–471, 2009. DOI: 10.1016/j.ultrasmedbio.2008.09.011. 2

[36] J.A. Noble, N. Navab, and H. Becher, "Ultrasonic image analysis and image quided interventions," *Interface Focus*, vol. 1, pp. 673–685, 2011. DOI: 10.1098/rsfs.2011.0025. 2

[37] D.T. Kuan and A.A. Sawchuk, "Adaptive noise smoothing filter for images with signal dependent noise," *IEEE Trans. Pattern Analysis & Machine Intelligence*, vol. PAMI-7, no. 2, pp. 165–177, 1985. DOI: 10.1109/TPAMI.1985.4767641. 119

[38] M. Nagao and T. Matsuyama, "Edge preserving smoothing," *Comput. Graph. & Image Proces.*, vol. 9, pp. 394–407, 1979. DOI: 10.1016/0146-664X(79)90102-3. 119

[39] T. Huang, G. Yang, and G. Tang, "A fast two-dimensional median filtering algorithm," *IEEE Trans. Acoustics, Speech & Signal Proces.*, vol. 27, no. 1, pp. 13–18, 1979. DOI: 10.1109/TASSP.1979.1163188.

[40] R. Gonzalez and R. Woods, Digital Image Processing, 2nd ed., Prentice-Hall, 2002. 80

[41] S. Solbo and T. Eltoft, "Homomorphic wavelet based-statistical despeckling of SAR images," *IEEE Trans. Geosc. Remote Sensing*, vol. 42, no. 4, pp. 711–721, 2004. DOI: 10.1109/TGRS.2003.821885. 70, 71, 72, 73, 74

[42] J. Saniie, T. Wang, and N. Bilgutay, "Analysis of homomorphic processing for ultrasonic grain signal characterization," *IEEE Trans. Ultras., Ferroel. & Frequ. Contr.*, vol. 3, pp. 365–375, 1989. DOI: 10.1109/58.19177. 73

[43] A. Nieminen, P. Heinonen, and Y. Neuvo, "A new class of detail-preserving filters for image processing," *IEEE Trans. Pattern Anal. Mach. Intell.*, vol. 9, pp. 74–90, 1987. DOI: 10.1109/TPAMI.1987.4767873.

[44] M. Kuwahara, K. Hachimura, S. Eiho, and M. Kinoshita, "Digital processing of biomedical images," Plenum. Pub. Corp., Eds. K. Preston and M. Onoe, pp. 187–203, 1976.

[45] A. Buades, B. Coll, and J.-M. Morel, "Nonlocal image and movie denoising," *Int. J. Comput. Vis.*, vol. 76, pp. 123–139, 2008. DOI: 10.1007/s11263-007-0052-1. 5

[46] S. Jin, Y. Wang, and J. Hiller, "An adaptive non-linear diffusion algorithm for filtering medical images," *IEEE Trans. Information Technology in Biomedicine*, vol. 4, no. 4, pp. 298–305, Dec. 2000. DOI: 10.1109/4233.897062. 39, 70, 71, 72, 73, 110, 128

[47] J. Weickert, B. Romery, and M. Viergever, "Efficient and reliable schemes for nonlinear diffusion filtering," *IEEE Trans. Image Processing*, vol. 7, pp. 398–410, 1998. DOI: 10.1109/83.661190. 72

[48] N. Rougon and F. Preteux, "Controlled anisotropic diffusion," *Conference on Nonlinear Image Processing VI*, IS&T/SPIE Symposium on Electronic Imaging, Science and Technology, San Jose, California, pp. 1–12, 5–10 Feb. 1995. 71, 79, 128

[49] M. Black, G. Sapiro, D. Marimont, and D. Heeger, "Robust anisotropic diffusion," *IEEE Trans. Image Processing*, vol. 7, no. 3, pp. 421–432, March 1998. DOI: 10.1109/83.661192. 71

[50] P. Rerona and J. Malik, "Scale-space and edge detection using anisotropic diffusion," *IEEE Trans. Pattern Analysis and Mach. Intelligence*, vol. 12, no. 7, pp. 629–639, July 1990. DOI: 10.1109/34.56205. 70, 74

[51] K. Abd-Elmoniem, A.-B. Youssef, and Y. Kadah, "Real-time speckle reduction and coherence enhancement in ultrasound imaging via nonlinear anisotropic diffusion," *IEEE Trans. Biomed. Eng.*, vol. 49, no. 9, pp. 997–1014, Sept. 2002. DOI: 10.1109/TBME.2002.1028423. 51, 70, 71, 72, 73, 74, 80

[52] R. Bernardes, C. Maduro, P. Serranho, A. Araujo, S. Barbeiro, and J. Cunha-Vaz, "Improved adaptive complex diffusion despeckling filter," *Opt. Express*, vol. 18, pp. 24048–24059, 2010. DOI: 10.1364/OE.18.024048.

[53] S. Zhong and V. Cherkassky, "Image denoising using wavelet thresholding and model selection," *Proc. IEEE Int. Conf. Image Process.*, Vancouver, Canada, pp. 1–4, Nov. 2000. DOI: 10.1109/ICIP.2000.899365. 74

[54] A. Achim, A. Bezerianos, and P. Tsakalides, "Novel Bayesian multiscale method for speckle removal in medical ultrasound images," *IEEE Trans. Med. Imag.*, vol. 20, no. 8, pp. 772–783, 2001. DOI: 10.1109/42.938245. 70, 71, 73, 74

[55] X. Zong, A. Laine, and E. Geiser, "Speckle reduction and contrast enhancement of echocardiograms via multiscale nonlinear processing," *IEEE Trans. Med. Imag.*, vol. 17, no. 4, pp. 532–540, 1998. DOI: 10.1109/42.730398. 71, 72, 73, 74

[56] X. Hao, S. Gao, and X. Gao, "A novel multiscale nonlinear thresholding method for ultrasonic speckle suppressing," *IEEE Trans. Med. Imag.*, vol. 18, no. 9, pp. 787–794, 1999. DOI: 10.1109/42.802756. 70, 71, 72, 73, 74

[57] D.L. Donoho, "Denoising by Soft Thresholding," *IEEE Trans. Inform. Theory*, vol. 41, pp. 613–627, 1995. DOI: 10.1109/18.382009. 71, 74

[58] F.N.S. Medeiros, N.D.A. Mascarenhas, R.C.P Marques, and C.M. Laprano, "Edge preserving wavelet speckle filtering," *5th IEEE Southwest Symposium on Image Analysis and Interpretation*, Santa Fe, New Mexico, pp. 281–285, 7–9 April 2002. DOI: 10.1109/IAI.2002.999933.

[59] S. Finn, M. Glavin, and E. Jones, "Echocardiographic speckle reduction comparison," *IEEE Trans. Ultr. Fer. Freq. Contr.*, vol. 58, no. 1, pp. 82–101, 2011. DOI: 10.1109/TUFFC.2011.1776. 5, 74

[60] M. Zain, M. Luqman, I. Elamvazuthi, and K.M. Begam, "Enhancement of bone fracture image using filtering techniques," *Int. J. Video Image Process. Netw. Secur. (IJVIPNS)*, vol. 9, no. 10, pp. 49–54, 2009. 5

[61] H.G. Senel, R.A. Peters, and B. Dawant, "Topological median filter," *IEEE Trans. Image Process.*, vol. 11, no. 2, 2002. DOI: 10.1109/83.982817. 5

[62] Q. Sun, J.A. Hossack, J.S. Tang, and S.T. Acton, "Speckle reducing anisotropic diffusion for 3D ultrasound images," *Comput. Med. Imaging Graph.*, vol. 28, pp. 461–470, 2004. DOI: 10.1016/j.compmedimag.2004.08.001. 5, 32, 35

[63] Y.S. Kim and J.B. Ra, "Improvement of ultrasound image based on wavelet transform: speckle reduction and edge enhancement," *Proc. SPIE, Image Processing Medical Imaging*, 5747, 2005. DOI: 10.1117/12.595129 . 5

[64] E. Nadernejad, "Despeckle filtering in medical ultrasound imaging," *Contemp. Eng. Sci.*, vol. 2, no. 1, pp. 17–36, 2009. 5

[65] G. Amara, "An introduction to wavelets," *IEEE Comput. Sci. Eng.*, vol. 2, no. 2, pp. 50–61, 1995. DOI: 10.1109/99.388960. 9

[66] A. Pižurica, W. Philips, I. Lemahieu, and M. Acheroy, "A versatile wavelet domain noise filtration technique for medical imaging," *IEEE Trans. Med. Imaging*, vol. 22, no. 3, pp. 323–331, 2003. DOI: 10.1109/TMI.2003.809588.

[67] Y. Zhan, M. Ding, L. Wu, and X. Zhang, "Nonlocal means method using weight refining for despeckling of ultrasound images," *Sign. Proces.*, vol. 103, pp. 201–213, 2014. DOI: 10.1016/j.sigpro.2013.12.019.

[68] M. Maggioni, G. Boracchi, A. Foi, and K. Engiazarian, "Video denoising, deblocking and enhancement through separable 4-D nonlocal spatiotemporal transforms," *IEEE Trans. Imag. Proc.*, vol. 21, no. 9, pp. 3952–3966, 2012. DOI: 10.1109/TIP.2012.2199324. 71, 86

[69] C. Lui and W.T. Freeman, "A high-quality video denoising algorithm based on reliable motion estimation," *European Conf. on Computer Vision*, pp. 706–719, 2010. DOI: 10.1007/978-3-642-15558-1_51.

[70] T.W. Chan, O.C. Au, T.S. Chong, and W.S. Chau, "A novel content-adaptive video denoising filter," *IEEE Proc. Vision, Image and Signal Proces.*, vol. 2, no. 2, pp. 649–652, 2005. DOI: 10.1109/ICASSP.2005.1415488.

[71] G. Varghese and Z. Wang, "Video denoising using a spatiotemporal statistical model of wavelet coefficients," *IEEE Int. Conf. on Acoustic, Speech and Sign. Proc.*, pp. 1257–1260, 2008. DOI: 10.1109/ICASSP.2008.4517845. 71

[72] N. Biradar, M.L. Dewal, and M.K. Rohit, "Edge preserved speckle noise reduction using integrated fuzzy filters," *Int. Scholarl. Research. Notic.*, vol. 2014, pp. 1–14, 2014. DOI: 10.1155/2014/876434.

[73] P. Coupé, P. Hellier, C. Kervrann, and C. Barillot, "Nonlocal means-based speckle filtering for ultrasound images," *IEEE Trans. on Image Proces.*, vol. 18, no. 10, pp. 2221–2229, 2009. DOI: 10.1109/TIP.2009.2024064.

[74] V. Zlokolica, W. Philips, and van de Ville, "Robust non-linear filtering for video process-ing," *IEEE Proc. Vision, Image and Signal Proces.*, vol. 2, no. 2, pp. 571–574, 2002. DOI: 10.1109/ICDSP.2002.1028154. 71, 85, 86

[75] V. Zlokolica, A. Pizurica, and W. Philips, "Recursive temporal denoising and motion es-timation of video," *Int. Conf. on Image Processing*, vol. 3 no. 3, pp. 1465–1468, 2008. DOI: 10.1109/ICIP.2004.1421340. 71, 85

[76] T. Loupas, W.N. McDicken, and P.L. Allan, "An adaptive weighted median filter for speckle suppression in Medical ultrasonic images," *IEEE Trans. Circuits and Systems*, vol. 36, pp. 129–135, 1989. DOI: 10.1109/31.16577. 80

[77] F.M. Cardoso, M.M. Matsumoto, and S.S. Furuie, "Edge-preserving speckle texture re-moval by interference-based speckle filtering followed by anisotropic diffusion," *Ultrasound Med. Biol.*, vol. 38, no. 8, pp. 1414–28, 2012. DOI: 10.1016/j.ultrasmedbio.2012.03.014.

[78] S.K. Narayanan and R.S.D. Wahidabanu, "Despeckling of ultrasound images, using me-dian regularized coupled Pde," *ACEEE Int. J. Contr. Syst. Instrum.*, vol. 2, no. 2, 2011.

[79] J.S. Ullom, M.L. Oelze, and J.R. Sanchez, "Speckle reduction for ultrasonic imag-ing using frequency compounding and despeckling filters along with coded excitation and pulse compression," *Advanc. Acoustics Vibration*, vol. 2012, pp. 1–16, 2012. DOI: 10.1155/2012/474039.

[80] P.C. Tay, C.D. Garson, S.T. Acton, and J.A. Hossack, "Ultrasound despeckling for contrast enhancement," *IEEE Trans. Image Proces.*, vol. 19, no. 7, pp. 1847–1860, 2010. DOI: 10.1109/TIP.2010.2044962. 122

[81] G. Ramos-Llorden, G. Vegas-Sanchez-Ferrero, M. Martin-Fernandez, C. Alberola-Lopez, and S. Aia-Fernandez, "Anisotropic diffusion filter with memory based on speckle statistics for ultrasound images," *IEEE Trans. Imag. Proces.*, vol. 24, no. 1, pp. 345–358, 2015. DOI: 10.1109/TIP.2014.2371244.

[82] B.A. Abrahim, Z.A. Mustafa, I.A. Yassine, N. Zayed, and Y.M. Kadah, "Hybrid total variation and wavelet thresholding speckle reduction for medical ultrasound imaging," *J. Med. Imag. and Health Inform.*, vol. 2, pp. 114–124, 2012. DOI: 10.1166/jmihi.2012.1072.

[83] J.M. Bioucas-Dias and M.A.T. Figueiredo, "Multiplicative noise removal using vari-able splitting and constrained optimization," *IEEE Trans. Imag. Proces.*, vol. 9, no. 10, pp. 1720–1730, 2010. DOI: 10.1109/TIP.2010.2045029.

[84] K. Dabov, A. Foi, and K. Egiazarian, "Video denoising by sparse 3D transform-domain collaborative filtering," *Proc. of the 15th Eur. Sign. Proc. Conf.*, pp. 1–5, 2007.

[85] D. Rusanovskyy, K. Dabov, and K. Egiazarian, "Moving-window varying size 3D transform-based video denoising," *Proc. Int. Workshop Video Proc. Quality Metrics*, pp. 1–4, 2006.

[86] S. Gupta, R.C. Chauhan, and S.C. Sexana, "Wavelet-based statistical approach for speckle reduction in medical ultrasound images," *Medical Biolog. Engineer. & Computing*, vol. 42, pp. 189–192, 2004. DOI: 10.1007/BF02344630.

[87] Y. Yue, M.M. Croitoru, A. Bidani, J.B. Zwischenberger, and J.W. Clark, Jr., "Nonlinear multiscale wavelet diffusion for speckle suppression and edge enhancement in ultrasound images," *IEEE Trans. Med. Imaging*, vol. 25, no. 3, pp. 297–311, Mar. 2006. DOI: 10.1109/TMI.2005.862737.

[88] R. Sivakumar, M.K. Gayathri, and D. Nedumaran, "Speckle filtering of ultrasound B-scan images-A comparative study of single scale spatial adaptive filters, multiscale filter and diffusion filters," *IACSIT Int. J. Engin. Techn.*, vol. 2, no. 6, pp. 514–523, 2010. DOI: 10.7763/IJET.2010.V2.174.

[89] J. Zhang, C. Wang, and Y. Cheng, "Comparison of despeckle filters for breast ultrasound images," *Circuits Syst. Process.*, vol. 34, pp. 185–208, 2015. DOI: 10.1007/s00034-014-9829-y.

[90] S.H. Contrera Ortiz, T. Chiu, and M.D. Fox, "Ultrasound image enhancement: A review," *Biomed. Sign. Proces. Contr.*, pp. 419–428, 2012. DOI: 10.1016/j.bspc.2012.02.002.

[91] N. Biradar, M.L. Dewal, and M.K. Rohit, "Comparative analysis of despeckle filters for continuous wave Doppler images," *Biomed. Eng. Letters*, vol. 5, no. 1, pp. 33–44, 2015. DOI: 10.1007/s13534-015-0171-5.

[92] P.H. Davis, J.D. Dawson, M.B. Biecha, R.K. Mastbergen, and M. Sonka, "Measurement of aortic intimal-media thickness in adolescents and young adults," *Ultr. Med. Biol.*, vol. 36, no. 4, pp. 560–565, 2010. DOI: 10.1016/j.ultrasmedbio.2010.01.002. 9

[93] X.G. Xu, Y.H. Na, and T. Zhang, "Design and test of a PC-based 3D ultrasound software system Ultra3D," *Comp. Biol. Med.*, vol. 38, no. 2, pp. 244–251, 2008. DOI: 10.1016/j.compbiomed.2007.10.003. 9

[94] E. Heiberg, J. Sjogren, M. Ugander, M. Carlsson, H. Engblom, and K. Arheden, "Design and validation of a segment-freely available software for cardiovascular image analysis," *BMC Med. Imag.*, vol. 10, no. 1, pp. 1–13, 2010. DOI: 10.1186/1471-2342-10-1. 9

[95] E.C. Kyriacou, S. Petroudi, C.S. Pattichis, M.S. Pattichis, M.B. Griffin, S. Kakkos, and A. Nicolaides, "Prediction of high risk asymptomatic carotid plaques based on ultrasonic image features," *IEEE Trans. Information Techn. Biomed.*, vol. 16, no. 5, pp. 966–973, 2012. DOI: 10.1109/TITB.2012.2192446. 9

[96] R. Cardenes-Almeida, A. Tristan-Vega, G.V.-S. Ferrero, S.A.-Fernandez, et al., "Usimag-tool: an open source freeware software for ultrasound imaging and electrography," *Int. Work. Multimodal Interfaces, eNTERFACE*, Istanbul, Turkey, pp. 117–127, 2007.

[97] F. Molinari, K.M. Meiburger, and J. Suri, "Automated high-performance cIMT measurement techniques using patented AtheroEdge™: A screening and home monitoring system," *33rd An. Int. Conf. IEEE EMBS*, Boston, pp. 6651–6654, 2011. DOI: 10.1109/IEMBS.2011.6091640. 9

[98] C.P. Loizou, C.S. Pattichis, M. Pantziaris, T. Tyllis, and A. Nicolaides, "Snakes based segmentation of the common carotid artery intima media," *Med. Biol. Eng. Comput.*, vol. 45, no. 1, pp. 35–49, Jan. 2007. DOI: 10.1007/s11517-006-0140-3. 11, 15, 17, 18, 19, 20, 21, 22, 23, 39, 53, 73, 79, 80, 120

[99] C.P. Loizou, C.S. Pattichis, A. Nicolaides, and M. Pantziaris, "Manual and automated media and intima thickness measurements of the common carotid artery," *IEEE Trans. Ultras. Ferroel. Freq. Contr.*, vol. 56, no. 5, pp. 983–994, 2009. DOI: 10.1109/TUFFC.2009.1130. 15, 16, 18, 19

[100] C.I. Christodoulou, S.C. Michaelides, and C.S. Pattichis, "Multi-feature texture analysis for the classification of clouds in satellite imagery," *IEEE Trans. Geoscience & Remote Sens.*, vol. 41, no. 11, pp. 2662–2668, Nov. 2003. DOI: 10.1109/TGRS.2003.815404. 70, 86, 128

[101] C.P. Loizou, M. Pantziaris, M.S. Pattichis, E. Kyriakou, and C.S. Pattichis, "Ultrasound image texture analysis of the intima and media layers of the common carotid artery and its correlation with age and gender," *Comput. Med. Imag. Graph.*, vol. 33, no. 4, pp. 317–324, 2009. DOI: 10.1016/j.compmedimag.2009.02.005. 15, 16, 18

[102] M.L. Bots, A.W. Hoes, P.J. Koudstaal, A. Hofman A, and D.E. Grobbee, "Common carotid intima-media thickness and risk of stroke and myocardial infarction: the Rotterdam Study," *Circulation*, vol. 96, pp. 1432–1437, 1997. DOI: 10.1161/01.CIR.96.5.1432. 15, 17, 128

[103] J.E. Wilhjelm, M.L. Gronholdt, B. Wiebe, S.K. Jespersen, L.K. Hansen, and H. Sillesen, "Quantitative analysis of ultrasound B-mode images of carotid atherosclerotic plaque: Correlation with visual classification and histological examination," *IEEE Trans. Med Imag.*, vol. 17, no. 6, pp. 910–922, 1998. DOI: 10.1109/42.746624. 15, 17

[104] D. Lamont, L. Parker, M. White, N. Unwin, et al., "Risk of cardiovascular disease measured by carotid intima-media thickness at age 49–51: life course study," *Biomed. Journ.*, vol. 320, pp. 273–278, 2000. 15

[105] C.D. Mario, G. Gorge, R. Peters, F. Pinto, et al., "Clinical application and image interpretation in coronary ultrasound. Study group of intra-coronary imaging of the working group of coronary circulation and of the subgroup of intravascular ultrasound of the working group of echocardiography of the European Society of Cardiology," *Eur. Heart. J.*, vol. 19, pp. 201–229, 1998. 16

[106] M.L. Grønhold, B.G. Nordestgaard, T.V. Schroeder, S. Vorstrup, and H. Sillensen, "Ultrasonic echolucent carotid plaques predict future strokes," *Circulation*, vol. 104, no. 1, pp. 168–173, 2001. DOI: 10.1016/S1567-5688(01)80045-3. 16

[107] E.J. Gussenhoven, P.A. Frietman, S.H. The, R.J. van Suylen, F.C. van Egmond, C.T. Lancee, H. van Urk, J.R. Roelandt, T. Stijnen, and N. Bom, "Assessment of medial thinning in atherosclerosis by intravascular ultrasound," *Am. J. Cardiol.*, vol. 68, pp. 1625–1632, 1991. DOI: 10.1016/0002-9149(91)90320-K. 16

[108] I. Wendendelhag, Q. Liang, T. Gustavsson, and J. Wikstrand, "A new automated computerized analysing system simplifies reading and reduces the variability in ultrasound measurement of intima media thickness," *Stroke*, vol. 28, pp. 2195–2200, 1997. DOI: 10.1161/01.STR.28.11.2195. 17, 39

[109] C.K. Zarins, C. Xu, and S. Glagov, "Atherosclerotic enlargement of the human abdominal aorta," *Atherosclerosis*, vol. 155, no. 1, pp. 157–164, 2001. DOI: 10.1016/S0021-9150(00)00527-X. 17

[110] ACAS clinical advisory, "Carotid endarterectomy for patients with asymptomatic internal carotid artery stenosis," *Stroke*, vol. 25, no. 12, pp. 2523–2524, 1994. DOI: 10.1161/01.STR.25.12.2523. 17, 30

[111] Executive Committee for the Asymptomatic Carotid Atherosclerosis study, "Endarterectomy for asymptomatic carotid stenosis," *J. Am. Med. Assoc.*, vol. 273, pp. 1421–1428, 2002. DOI: 10.1001/jama.1995.03520420037035. 17, 31

[112] A.N. Nicolaides, M. Sabetai, S.K. Kakkos, S. Dhanjil, T. Tegos, and J.M. Stevens, "The Asymptomatic carotid stenosis and risk of stroke study," *Int. Angiol.*, vol. 22, no. 3, pp. 263–272, 2003. 17, 31, 53, 86

[113] C. Metz, "Basic principles of ROC analysis," *Semin. Nucl. Medic.*, vol. 8, pp. 283–298, 1978. DOI: 10.1016/S0001-2998(78)80014-2. 34, 36

[114] D. Williams and M. Shah, "A fast algorithm for active contour and curvature estimation," *GVCIP: Imag. Und.*, vol. 55, no. 1, pp. 14–26, 1992. DOI: 10.1016/1049-9660(92)90003-L. 18, 21, 31, 34, 35, 50

[115] V. Chalana and Y. Kim, "A methodology for evaluation of boundary detection algorithms on medical images," *IEEE Trans. Med. Imaging*, vol. 16, no. 5, pp. 642–652, 1997. DOI: 10.1109/42.640755. 23

[116] D.C. Cheng, A. Schmidt-Trucksass, K.S. Cheng, and H. Burkhardt, "Using snakes to detect the intimal and adventitial layers of the common carotid artery wall in sonographic images," *Comput. Methods Programs Biomed.*, vol. 67, no. 1, pp. 27–37, Jan. 2002. DOI: 10.1016/S0169-2607(00)00149-8. 23, 31

[117] D.E. Ilea, C. Duffy, L. Kavanagh, A. Stanton, and P.F. Whelan, "Fully automated segmentation and tracking of the intima media thickness in ultrasound video sequences of the common carotid artery," *IEEE Trans. Ultr. Fer. & Freq. Contr.*, vol. 60, no. 1, pp. 158–177, 2013. DOI: 10.1109/TUFFC.2013.2547. 25, 30

[118] C.P. Loizou, T. Kasparis, P. Papakyriakou, L. Christodoulou, M. Pantziaris, and C.S. Pattichis, "Video segmentation of the common carotid artery intima-media complex," *12th Int. Conf. Bioinf. & Bioeng. Proc. (BIBE)*, Larnaca, Cyprus, Nov. 11–13, pp. 500–505, 2012. DOI: 10.1109/BIBE.2012.6399728. 20, 27, 28, 29, 30

[119] R.H. Selzer, W.J. Mack, P.L. Lee, H. Kwong-Fu, and H.N. Hodis, "Improved carotid elasticity and intima-media thickness measurements from computer analysis of sequential ultrasound frames," *Atherosclerosis*, vol. 154, no. 1, pp. 185–193, 2001. DOI: 10.1016/S0021-9150(00)00461-5. 29

[120] F. Destrempes, J. Meunier, M.-F. Giroux, G. Soulez, and G. Cloutier, "Segmentation of plaques in sequences of ultrasonic B-mode images of carotid based on motion estimation and a Bayesian model," *IEEE Trans. Biomed. Eng.*, vol. 58, no. 8, pp. 2202–2211, 2011. DOI: 10.1109/TBME.2011.2127476. 30, 34, 40, 51

[121] K.F. Lai and R.T. Chin, "Deformable contours-modelling and extraction," *IEEE Trans. on PAMI*, vol. 17, no. 11, pp. 1084–1090, 1995. DOI: 10.1109/34.473235. 31, 34

[122] C. Xu and J. Prince, "Generalized Gradient vector flow external forces for active contours," *Signal Processing*, vol. 71, pp. 131–139, 1998. DOI: 10.1016/S0165-1684(98)00140-6. 31, 34

[123] A. Hamou and M. El.-Sakka, "A novel segmentation technique for carotid ultrasound images," *Int. Conf. on Acoustic Speech and Signal Processing (ICASSP)*, pp. III-521–III-524, 2004. DOI: 10.1109/ICASSP.2004.1326596. 50, 53

[124] A.R. Abdel-Dayen and M.R. El.-Sakka, "A novel morphological-based carotid artery contour extraction," *Canadian Conf. Electr. and Comp. Engin.*, vol. 4, pp. 1873–1876, 2–5 May 2004. DOI: 10.1109/CCECE.2004.1347574. 50

[125] F. Mao, J. Gill, D. Downey, and A. Fenster, "Segmentation of carotid artery in ultrasound images: Method development and evaluation technique," *Med. Phys.*, vol. 27, no. 8, pp. 1–10, 2000. DOI: 10.1118/1.1287111. 50

[126] P. Abolmaesumi, M.R. Sirouspour, and S.E. Salcudean, "Real-time extraction of carotid artery contours from ultrasound images," *Proc. IEEE Int. Conf. Computer Based Medical Systems*, pp. 181–186, 2000. DOI: 10.1109/CBMS.2000.856897. 50

[127] L.D. Cohen, "On active contour models and balloons," *Comp. Vis., Graph., and Imag. Proces.: Imag. Underst. (CVGIP:IU)*, vol. 53, no. 2, pp. 211–218, 1991. DOI: 10.1016/1049-9660(91)90028-N. 34, 50

[128] J.D. Gill, H.M. Ladak, D.A. Steinman, and A. Fenster, "Segmentation of ulcerated plaque: A semi-automatic method for tracking the progression of carotid atherosclerosis," *World congress on Med. Phys. and Biomed. Eng.*, Chicago, IL, pp. 1–4, 2000. DOI: 10.1109/IEMBS.2000.900833. 50

[129] S. Delsanto, F. Molinari, P. Giustetto, W. Liboni, S. Badalamenti, and J.S. Suri, "Characterization of completely user-independent algorithm for carotid artery segmentation in 2D ultrasound images," *IEEE Trans. Instr. Meas.*, vol. 56, no. 4, pp. 1265–1274, 2007. DOI: 10.1109/TIM.2007.900433. 51

[130] S. Golemati, J. Stoitsis, E.G. Sifakis, T. Balkisas, and K.S. Nikita, "Using the Hough transform to segment ultrasound images of longitudinal and transverse sections of the carotid artery," *Ultrasound Med. Biol.*, vol. 33, no. 12, pp. 1918–1932, 2007. DOI: 10.1016/j.ultrasmedbio.2007.05.021. 48, 51

[131] A. Zahalka and A. Fenster, "An automated segmentation method for three-dimensional carotid ultrasound images," *Phys. Med. Biol.*, vol. 46, no. 4, pp. 1321–1342, 2001. DOI: 10.1088/0031-9155/46/4/327. 51

[132] E. Ukwatta, J. Yuan, D. Buchanan, B. Chiu, J. Awad, W. Qiu, G. Parraga, and A. Fenster, "Three-dimensional segmentation of three-dimensional ultrasound carotid atherosclerosis using sparse field level sets," *Med. Phys.*, vol. 40, no. 5, 17 pages, 2013. DOI: 10.1118/1.4800797. 51

[133] J.H. Dwyer, P. Sun, H. Kwong-Fu, K.M. Dwyer, and R.H. Selzer, "Automated intima-media thickness: The Los Angeles atherosclerosis study," *Ultrasound Med. Biol.*, vol. 24, no. 7, pp. 981–987. 1998. DOI: 10.1016/S0301-5629(98)00069-6.

[134] I. Wendelhag, T. Gustavsson, M. Suurkula, G. Berglund, and J. Wikstrand, "Ultrasound measurement of wall thickness in the carotid artery: Fundamental principles and description of a computerized analyzing system," *Clin. Physiol.*, vol. 11, no. 6, pp. 565–577, 1991. DOI: 10.1111/j.1475-097X.1991.tb00676.x. 19

[135] C. Liguori, A. Paolillo, and A. Pietrosanto, "An automatic measurement system for the evaluation of carotid intima-media thickness," *IEEE Trans. Instrum. Meas.*, vol. 50, no. 6, pp. 1684–1691, 2001. DOI: 10.1109/19.982968.

[136] Q. Liang, I. Wendelhag, J. Wikstrand, and T. Gustavsson, "A multiscale dynamic programming procedure for boundary detection in ultrasonic artery images," *IEEE Trans. Med. Imag.*, vol. 19, no. 2, pp. 127–142, Feb. 2000. DOI: 10.1109/42.836372. 39

[137] F. Faita, V. Gemignani, E. Bianchini, C. Giannarelli, and M. Demi, "Real-time measurement system for the evaluation of the intima media thickness with a new edge detector," in *Proc. 28th Ann. Int.Conf. IEEE Eng. Med. Biol. Soc.*, pp. 715–718, 2006. DOI: 10.1109/IEMBS.2006.260450.

[138] M.A. Gutierrez, P.E. Pilon, S.G. Lage, L. Kopel, R.T. Carvalho, and S.S. Furuie, "Automatic measurement of carotid diameter and wall thickness in ultrasound images," *Comput. Cardiol.*, vol. 29, pp. 359–362, 2002. DOI: 10.1109/CIC.2002.1166783.

[139] R.C. Chan, J. Kaufhold, L.C. Hemphill, R.S. Lees, and W.C. Karl, "Anisotropic edge-preserving smoothing in carotid B-mode ultrasound for improved segmentation and intima-media thickness (IMT) measurement," *Comput. Cardiol.*, vol. 27, pp. 37–40, 2000. DOI: 10.1109/CIC.2000.898449.

[140] R. Rocha, A. Campilho, J. Silva, E. Azevedo, and R. Santos, "Segmentation of the carotid intima-media region in B-mode ultrasound images," *Image Vis. Comput.*, vol. 28, no. 4, pp. 614–625, Apr. 2010. DOI: 10.1016/j.imavis.2009.09.017.

[141] F. Molinari, G. Zeng, and J.S. Suri, "Intima-media thickness: Setting a standard for a completely automated method of ultrasound measurement," *IEEE Trans. Ultrason. Ferroelectr. Freq. Control*, vol. 57, no. 5, pp. 1112–1124, 2010. DOI: 10.1109/TUFFC.2010.1522.

[142] F. Molinari, C. Pattichis, G. Zeng, L. Saba, U. Acharya, R. Sanfilippo, A. Nicolaides, and J. Suri, "Completely automated multiresolution edge snapper (CAMES)—A new technique for an accurate carotid ultrasound IMT measurement: Clinical validation and benchmarking on a multi-institutional database," *IEEE Trans. Image Process.*, vol. 21, no. 3, pp. 1211–1222, 2012. DOI: 10.1109/TIP.2011.2169270.

[143] S. Petroudi, C.P. Loizou, M. Pantziaris, and C.S. Pattichis, "Segmentation of the common carotid intima-media complex in ultrasound images using active contours," *IEEE Trans. Biomed. Eng.*, vol. 59, no. 11, pp. 3060–3069, 2012. DOI: 10.1109/TBME.2012.2214387. 30, 39, 40

[144] C.K. Zarins, C. Xu, and S. Glagov, "Atherosclerotic enlargement of the human abdominal aorta," *Atherosclerosis*, vol. 155, no. 1, pp. 157–164, 2001. DOI: 10.1016/S0021-9150(00)00527-X. 39, 70

[145] F.J. Polak, Doppler Sonography: An Overview, In Peripheral Vascular Sonography: A Practical Guide, Baltimore: Williams and Wilkins, 1992. 39

[146] D. Kashiwazaki, T. Yoshimoto, T. Mikami, M. Muraki, S. Fujimoto, K. Abiko, and S. Kaneko, "Identification of high-risk carotid artery stenosis: motion of intra plaque contents detected using B-mode ultrasonography," *J. of Neurosurgery*, vol. 117, no. 3, pp. 574–578, 2012. DOI: 10.3171/2012.6.JNS111922. 40

[147] C.S. Pattichis, C. Christodoulou, E. Kyriakou, M. Pantziaris, A. Nicolaides, M.S. Pattichis, and C.P. Loizou, "Ultrasound imaging of carotid atherosclerosis," in *Wiley encyclopedia of Biomedical Engineering*, Eds. M. Akay, Wiley, Hoboken: John Wiley & Sons, 2006. 86, 120

[148] N. Liasis, C. Klonaris, A. Katsargyris, S. Georgopoulos, et al., "The use of speckle reduction imaging (SRI) ultrasound in the characterization of carotid artery plaques," *Eur. J. Radiol.*, vol. 65, pp. 427–433, 2008. DOI: 10.1016/j.ejrad.2007.05.004. 53

[149] V. Damerjian, O. Tankyevych, N. Souag, and E. Petit, "Speckle characterization in ultrasound images-A review," *IRBM*, vol. 35, pp. 202–213, 2014. DOI: 10.1016/j.irbm.2014.05.003. 53

[150] E. Brusseau, C.L. De Korte, F. Mastick, J. Schaar, and A.F.W. Van der Steen, "Fully automatic luminal contour segmentation in intracoronary ultrasound imaging-A statistical approach," *IEEE Trans. Med. Imag.*, vol. 23, no. 5, pp. 554–566, 2004. DOI: 10.1109/TMI.2004.825602. 80

[151] M.R. Cardinal, J. Meunier, G. Soulez, E. Thérasse, and G. Cloutier, "Intravascular Ultrasound Image Segmentation: A Fast-Marching Method," *Proc. MICCAI, LNCS 2879*, pp. 432–439, 2003. DOI: 10.1007/978-3-540-39903-2_53. 80

[152] R.M. Haralick, K. Shanmugam, and I. Dinstein, "Texture Features for Image Classification," *IEEE Trans. Systems, Man., and Cybernetics*, vol. SMC-3, pp. 610–621, Nov. 1973. DOI: 10.1109/TSMC.1973.4309314. 56, 70, 80

[153] J.S. Weszka., C.R. Dyer, and A. Rosenfield, "A comparative study of texture measures for terrain classification," *IEEE Trans. Systems, Man. & Cybernetics*, vol. SMC-6, pp. 269–285, April 1976. DOI: 10.1109/TSMC.1976.5408777. 56, 70

[154] F. Rakebrandt et al., "Relation between ultrasound texture classification images and histology of atherosclerotic plaque," *Journal of Ultrasound in Med. & Biology*, vol. 26, no. 9, pp. 1393–1402, 2000. DOI: 10.1016/S0301-5629(00)00314-8. 70

[155] M. Pattichis, C. Pattichis, M. Avraam, A. Bovik, and K. Kyriakou, "AM-FM texture segmentation in electron microscopic muscle imaging," *IEEE Trans. Med. Imag.*, vol. 19, no. 12, pp. 1253–1258, 2000. DOI: 10.1109/42.897818. 70

[156] N. Mudigonda, R. Rangayyan, and J. Desautels, "Detection of breast masses in mammograms by density slicing and texture flow-field analysis," *IEEE Trans. Med. Imag.*, vol. 20, no. 12, pp. 121–1227, 2001. DOI: 10.1109/42.974917. 70

[157] S.M. Ali and R.E. Burge, "New automatic techniques for smoothing and segmenting SAR images," *Signal Processing*, North-Holland, vol. 14, pp. 335–346, 1988. DOI: 10.1016/0165-1684(88)90092-8. 70

[158] A. Baraldi and F. Pannigianni, "A refined gamma MAP SAR speckle filter with improved geometrical adaptivity," *IEEE Trans. Geoscience and Remote Sensing*, vol. 33, no. 5, pp. 1245–1257, Sep. 1995. DOI: 10.1109/36.469489. 70, 71, 72, 73, 74

[159] R.N. Czerwinski, D.L. Jones, and W.D. O'Brien, "Detection and boundaries in speckle images-Application to medical ultrasound," *IEEE Trans. Med. Imag.*, vol. 18, no. 2, pp. 126–136, Feb. 1999. DOI: 10.1109/42.759114. 70

[160] M. Karaman, M. Alper Kutay, and G. Bozdagi, "An adaptive speckle suppression filter for medical ultrasonic imaging," *IEEE Trans. Med. Imag.*, vol. 14, no. 2, pp. 283–292, 1995. DOI: 10.1109/42.387710. 71, 73

[161] A.M. Wink and J.B.T.M. Roerdink, "Denoising functional MR images: A comparison of wavelet denoising and Gaussian smoothing," *IEEE Trans. Med. Imag.*, vol. 23, no. 3, pp. 374–387, 2004. DOI: 10.1109/TMI.2004.824234. 71

[162] H.-L. Eng and K.-K. Ma, "Noise adaptive soft-switching median filter," *IEEE Trans. Image Process.*, vol. 10, no. 2, pp. 242–251, 2001. DOI: 10.1109/83.902289. 71, 73, 74

[163] C.P. Loizou, C. Christodoulou, C.S. Pattichis, R. Istepanian, M. Pantziaris, and A. Nicolaides, "Speckle reduction in ultrasound images of atherosclerotic carotid plaque," *DSP 2002, Proc. IEEE 14th Int. Conf. Digital Signal Proces.*, Santorini-Greece, pp. 525–528, July 1–3, 2002. 72

[164] E. Trouve, Y. Chambenoit, N. Classeau, and P. Bolon, "Statistical and operational performance assessment of multi-temporal SAR image filtering," *IEEE Trans. Geosc. Remote Sens.*, vol. 41, no. 11, pp. 2519–2539, 2003. DOI: 10.1109/TGRS.2003.817270. 63, 73

[165] D. Schilling and P.C. Cosman, "Image quality evaluation based on recognition times for fast browsing image applications," *IEEE Trans. on Multimedia*, vol. 4, no. 3, pp. 320–331, Sept. 2002. DOI: 10.1109/TMM.2002.802844. 73

[166] A. Pommert and K. Hoehne, "Evaluation of image quality in medical volume visualization: The state of the art," Takeyoshi Dohi and Ron Kikinis (Eds.): *Medical image computing and computer-assisted intervention, Proc. MICCAI, 2002, Part II, Lecture Notes in Computer Science 2489*, pp. 598–605, Springer Verlag, Berlin 2002. DOI: 10.1007/3-540-45787-9. 73

[167] J.E. Wilhjelm, M.S. Jensen, S.K. Jespersen, B. Sahl, and E. Falk, "Visual and quantitative evaluation of selected image combination schemes in ultrasound spatial compound scanning," *IEEE Trans. Med. Imag.*, vol. 23, no. 2, pp. 181–190, 2004. DOI: 10.1109/TMI.2003.822824. 73, 79

[168] M. Eckert, "Perceptual quality metrics applied to still image compression," Canon information systems research, Faculty of engineering, University of Technology, Sydney, Australia, pp. 1–26, 2002. DOI: 10.1016/S0165-1684(98)00124-8. 74

[169] J.C. Bamber and C. Daft, "Adaptive filtering for reduction of speckle in ultrasonic pulse-echo images," *Ultrasonic*, vol. 24, pp. 41–44, 1986. DOI: 10.1016/0041-624X(86)90072-7. 79

[170] J.T.M. Verhoeven and J.M. Thijssen, "Improvement of lesion detectability by speckle reduction filtering: A quantitative study," *Ultrasonic Imaging*, vol. 15, pp. 181–204, 1993. DOI: 10.1006/uimg.1993.1012. 79

[171] C. Kotropoulos and I. Pitas, "Optimum nonlinear signal detection and estimation in the presence of ultrasonic speckle," *Ultrasonic Imaging*, vol. 14, pp. 249–275, 1992. DOI: 10.1016/0161-7346(92)90066-5. 79

[172] M.E. Olszewski, A. Wahle, S.C. Vigmostad, and M. Sonka, "Multidimensional segmentation of coronary intravascular ultrasound images using knowledge-based methods," *Med. Imag.: Image Processing, Proc. SPIE*, 5747, pp. 496–504, 2005. DOI: 10.1117/12.595850. 80

[173] Z. Wang and A. Bovik, "A Universal quality index," *IEEE Signal Processing Letters*, vol. 9, no. 3, pp. 81–84, March 2002. DOI: 10.1109/97.995823. 80, 81

[174] E. Krupinski, H. Kundel, P. Judy, and C. Nodine, "The medical image perception society, key issues for image perception research," *Radiology*, vol. 209, pp. 611–612, 1998. DOI: 10.1148/radiology.209.3.9844649. 86

[175] A. Ahumada and C. Null, "Image quality: A multidimensional problem," in *Digital images and human vision*, A.B. Watson (Ed.), Bradford Press: Cambridge, Mass, pp. 141–148, 1993.

[176] G. Deffner, "Evaluation of display image quality: Experts vs. non-experts," *Symposium Society for Information and Display Digest*, vol. 25, pp. 475–478, 1994.

[177] H.R. Sheikh, A.C. Bovik, and G. de Veciana, "An information fidelity criterion for image quality assessment using natural scene statistics," *IEEE Trans. Image Proces.*, vol. 14, no. 12, pp. 2117–2128, Dec. 2005. DOI: 10.1109/TIP.2005.859389. 71, 80

[178] B. Fetics et al., "Enhancement of contrast echocardiography by image variability analysis," *IEEE Trans. Med. Imag.*, vol. 20, no. 11, pp. 1123–1130, Nov. 2001. DOI: 10.1109/42.963815.

[179] M. Oezkan, A. Erdem, M. Sezan, and A. Tekalp, "Efficient multi-frame Wiener restoration of blurred and noisy image sequences," *IEEE Trans. Image Proces.*, vol. 1, pp. 453–476, Oct. 1992. DOI: 10.1109/83.199916. 84

[180] P.M.B. Van Roosmalen, S.J.P Westen, R.L. Lagendijk, and J. Biemond, "Noise reduction for image sequences using an oriented pyramid threshold technique," *IEEE Int. conf. Image Processing*, vol. 1, pp. 375–378, 1996. DOI: 10.1109/ICIP.1996.559511. 84

[181] M. Vetterli and J. Kovacevic, Wavelets and sub band coding, Prentice Hall, 1995. 84

[182] S. Winkler, "Digital video quality," Vision models and metrics, John Wiley & Sons, 2005. 84, 121

[183] J.-H. Jung, K. Hong, and S. Yang, "Noise reduction using variance characteristics in noisy image sequence," on *Int. Conf. Consumer Electronics*, pp. 213–214, 8-12 Jan. 2005. DOI: 10.1109/ICCE.2005.1429793. 84

[184] M. Bertalmio, V. Caselles, and A. Pardo, "Movie Denoising by average of warped lines," *IEEE Trans. Image Processing*, vol. 16, no. 9, pp. 233–2347, 2007. DOI: 10.1109/TIP.2007.901821. 85

[185] B. Alp, P. Haavisto, T. Jarske, K. Oestaemoe, and Y. Neuro, "Median based algorithms for image sequence processing," *SPIE Visual Communications and Image Processing*, pp. 122–133, 1990. DOI: 10.1117/12.24175. 85, 121, 127

[186] A Philips Medical System Company, "Comparison of image clarity, SonoCT real-time compound imaging versus conventional 2D ultrasound imaging," *ATL Ultrasound, Report*, 2001. 19, 128

[187] C.P. Loizou, T. Kasparis, T. Lazarou, C.S. Pattichis, and M. Pantziaris, "Manual and automated intima-media thickness and diameter measurements of the common carotid artery in patients with renal failure disease," *Comput. Biol. & Medicine*, vol. 53, pp. 220–229, 2014. DOI: 10.1016/j.compbiomed.2014.08.003.

[188] A.S. Panayides, M.S. Pattichis, and C.S. Pattichis, "Mobile-Health Systems Use Diagnostically Driven Medical Video Technologies [Life Sciences]," *IEEE Signal Processing Magazine*, vol. 30, no. 6, pp. 163–172, Nov. 2013. DOI: 10.1109/MSP.2013.2276512. DOI: 10.1109/MSP.2013.2276512. 95, 103, 104

[189] A.S. Panayides, Z.C. Antoniou, and A.G. Constantinides, "An Overview of mHealth Medical Video Communication Systems," in Mobile Health: A Technology Road Map, Edition: Springer Series in Bio-/Neuroinformatics, Chapter: 26. Publisher: Springer, pp. 609–633, 2015. DOI: 10.1007/978-3-319-12817-7. 101, 102

[190] A. Panayides, M.S. Pattichis, C.S. Pattichis, C.N. Schizas, A. Spanias, and E.C. Kyriacou, "An Overview of Recent End-to-End Wireless Medical Video Telemedicine Systems using 3G," *in Proc. of IEEE EMBC'10*, Buenos Aires, Argentina, Aug. 31–Sep. 4, 2010. DOI: 10.1109/IEMBS.2010.5628076. 95

[191] A. Panayides, M.S. Pattichis, C.S. Pattichis, and A. Pitsillides, "A Tutorial for Emerging Wireless Medical Video Transmission Systems [Wireless Corner]," *IEEE Antennas & Propagation Magazine*, vol. 53, no. 2, pp. 202–213, April 2011. DOI: 10.1109/MAP.2011.5549369. 98

[192] A.S. Panayides, "Diagnostically Resilient Encoding, Wireless Transmission, and Quality Assessment of Medical Video," Ph.D. Dissertation, Department of Computer Science, University of Cyprus, Nicosia, 2011.

[193] A. Panayides, M.S. Pattichis, A.G. Constantinides, and C.S. Pattichis, "M-Health Medical Video Communication Systems: An Overview of Design Approaches and Recent Advances," *in Proc. of IEEE EMBC'13*, Osaka, Japan, pp. 7253–7256, Jul. 3–7, 2013. DOI: 10.1109/EMBC.2013.6611232. 95

[194] ITU-T, "Video codec for audiovisual services ar px64 kbit/s," *ITU-T Recommendation H.261*, Nov. 1990. 96

[195] ITU-T, "Information Technology – Generic Coding of moving pictures and associated audio information: video," *ITU-T Recommendation H.262*, Jul. 1995. 96

[196] ITU-T, "Video coding for low bitrate communication," *ITU-T Recommendation H.263*, Nov. 1995. DOI: 10.1109/35.556485. 96

[197] ITU-T, "Advanced video coding for generic audiovisual services," *ITU-T and ISO/IEC 14496–10 Recommendation H.264 (MPEG4-AVC)*, May 2003. 96

[198] T. Wiegand, G.J. Sullivan, G. Bjontegaard, and A. Luthra, "Overview of the H.264/AVC video coding standard," *IEEE Transactions on Circuits and Systems for Video Technology*, vol. 13, p. 560–576, July 2003. DOI: 10.1109/TCSVT.2003.815165. 96

[199] ITU-T, "H.265: High efficiency video coding," *ITU-T Recommendation H.265*, June 2013. 96, 97

[200] G.J. Sullivan, J.-R. Ohm, W.-J. Han, and T. Wiegand, "Overview of the High Efficiency Video Coding (HEVC) Standard," *IEEE Trans. Circuits and Systems for Video Tech.*, vol. 22, no. 12, pp. 1649–1668, Dec. 2012. DOI: 10.1109/TCSVT.2012.2221191. 96, 97, 108

[201] J.-R. Ohm, G.J. Sullivan, H. Schwarz, T.K. Tan, and T. Wiegand, "Comparison of the Coding Efficiency of Video Coding Standards—Including High Efficiency Video Coding (HEVC)," *IEEE Trans. Circuits and Systems for Video Tech.*, vol. 22, no. 12, pp. 1669–1684, Dec. 2012. DOI: 10.1109/TCSVT.2012.2221192. 97

[202] Rysavy Research, LLC, "Mobile Broadband Explosion: The 3GPP Wireless Evolution," Aug. 2013. Available: http://www.4gamericas.org/. 98

[203] ITU-R Rep. M.2134, "Requirements Related to Technical Performance for IMT-Advanced Radio Interface(s)," Nov. 2008. 98

[204] The Draft IEEE 802.16m System Description Document (SDD), IEEE 802.16 Broadband Wireless Access Working Group, Jul. 2008. 99

[205] S. Ahmadi, "An overview of next-generation mobile WiMAX technology," *Communications Magazine, IEEE*, vol. 47, no. 6, pp. 84–98, June 2009. DOI: 10.1109/MCOM.2009.5116805.

[206] I. Papapanagiotou, D. Toumpakaris, L. Jungwon, and M. Devetsikiotis, "A survey on next generation mobile WiMAX networks: objectives, features and technical challenges," *Communications Surveys & Tutorials, IEEE*, vol. 11, no. 4, pp. 3–18, Fourth Quarter 2009. DOI: 10.1109/SURV.2009.090402. 99

[207] A. Ghosh, R. Ratasuk, B. Mondal, N. Mangalvedhe, and T. Thomas, "LTE-advanced: next-generation wireless broadband technology [Invited Paper]," *Wireless Communications, IEEE*, vol. 17, no. 3, pp. 10–22, June 2010. DOI: 10.1109/MWC.2010.5490974. 100

[208] S. Parkvall, A. Furuskär, and E. Dahlman, "Evolution of LTE toward IMT-advanced," *Communications Magazine, IEEE*, vol. 49, no. 2, pp. 84–91, Feb. 2011. DOI: 10.1109/MCOM.2011.5706315. 100

[209] S.P. Rao, N.S. Jayant, M.E. Stachura, E. Astapova, and A. Pearson-Shaver, "Delivering Diagnostic Quality Video over Mobile Wireless Networks for Telemedicine," *International Journal of Telemedicine and Applications*, vol. 2009, Article ID 406753, 9 pages, 2009. DOI: 10.1155/2009/406753. 105, 106

[210] N. Tsapatsoulis, C. Loizou, and C. Pattichis, "Region of Interest Video Coding for Low bit-rate Transmission of Carotid Ultrasound Videos over 3G Wireless Networks," in *Proc. of 29th Annual Conference of the IEEE Engineering in Medicine and Biology*

Society, IEEE EMBC'07, Lyon, France, pp. 3717–3720, August 22–26, 2007. DOI: 10.1109/IEMBS.2007.4353139. 105

[211] M.G. Martini and C.T.E.R. Hewage, "Flexible Macroblock Ordering for Context-Aware Ultrasound Video Transmission over Mobile WiMAX," *International Journal of Telemedicine and Applications*, vol. 2010, Article ID 127519, 14 pages, 2010. DOI: 10.1155/2010/127519. 105, 106

[212] A. Panayides, M.S. Pattichis, C.S. Pattichis, C.P. Loizou, M. Pantziaris, and A. Pitsillides, "Atherosclerotic Plaque Ultrasound Video Encoding, Wireless Transmission, and Quality Assessment Using H.264," *IEEE Transactions on Information Technology in Biomedicine*, vol. 15, no. 3, pp. 387–397, May 2011. DOI: 10.1109/TITB.2011.2105882. 105, 106, 110, 111, 112, 113, 115

[213] A. Panayides , Z. Antoniou , V. Barberis , M. Pattichis , C. Pattichis, and E. Kyriacou, "Abdominal aortic aneurysm medical video transmission," *Proc. IEEE-EMBS Int. Conf. Biomed. Health Inf.*, pp. 679–682, Hong Kong, China, Jan. 2–7, 2012. DOI: 10.1109/BHI.2012.6211674. 106

[214] A. Panayides, Z. Antoniou, Y. Mylonas, M.S. Pattichis, A. Pitsillides, and C.S. Pattichis, "High-Resolution, Low-delay, and Error-resilient Medical Ultrasound Video Communication Using H.264/AVC Over Mobile WiMAX Networks," *IEEE Journal Biomed. and Health Informatics*, vol. 17, no. 3, pp. 619–628, May 2013. doi: 10.1109/TITB.2012.2232675. DOI: 10.1109/TITB.2012.2232675. 105, 106, 107

[215] S. Khire, S. Robertson, N. Jayant, E.A. Wood, M.A. Stachura, and T. Goksel, "Region-of-interest video coding for enabling surgical tele mentoring in low-bandwidth scenarios," in *Proc. of MILCOM2012*, pp. 1–6, Oct. 29—Nov. 1, 2012. DOI: 10.1109/MILCOM.2012.6415792. 105

[216] E. Cavero, A. Alesanco, L. Castro, J. Montoya, I. Lacambra, and J. Garcia, "SPIHT-Based Echocardiogram Compression: Clinical Evaluation and Recommendations of Use," *IEEE Journal of Biomedical and Health Informatics*, vol. 17, no. 1, pp. 103–112, Jan. 2013. DOI: 10.1109/TITB.2012.2227336. 105, 106

[217] E. Cavero, A. Alesanco, and J. Garcia, "Enhanced protocol for real time transmission of echocardiograms over wireless channels," *IEEE Transactions on Biomedical Engineering*, vol. 59, no. 11, pp. 3212–3220, Nov. 2012. DOI: 10.1109/TBME.2012.2207720. 105, 106

[218] C. Debono, B. Micallef, N. Philip, A. Alinejad, R. Istepanian, and N. Amso, "Cross Layer Design for Optimised Region of Interest of Ultrasound Video Data over Mobile WiMAX," *IEEE Transactions on Information Technology in Biomedicine*, vol. 16, no. 6, pp. 1007–1014, Nov. 2012. DOI: 10.1109/TITB.2012.2201164. 105, 106

[219] A. Alinejad, N. Philip, and R. Istepanian, "Cross Layer Ultrasound Video Streaming over Mobile WiMAX and HSUPA Networks," *IEEE Transactions on Information Technology in Biomedicine*, vol. 16, no. 1, pp. 31–39, Jan. 2012. DOI: 10.1109/TITB.2011.2154384. 106

[220] Y. Chu and A. Ganz, "A Mobile Teletrauma System using 3G Networks," *IEEE Transactions*, in *Information Technology in Biomedicine*, vol. 8, no. 4, pp. 456–462, December, 2004. DOI: 10.1109/TITB.2004.837893.

[221] S.A. Garawi, R.S.H. Istepanian, and M.A. Abu-Rgheff, "3G Wireless Communication for Mobile Robotic Tele-ultrasonography Systems," *IEEE Communications Magazine*, vol. 44, no. 4, pp. 91–96, April, 2006. DOI: 10.1109/MCOM.2006.1632654. 106

[222] R.S.H. Istepanian, N.Y. Philip, and M.G. Martini, "Medical QoS Provision based on Reinforcement Learning in Ultrasound Streaming over 3.5G Wireless Systems," *IEEE Journal on Selected Areas in Communications*, vol. 27, no. 4, pp. 566–574, May 2009. DOI: 10.1109/JSAC.2009.090517. 106

[223] P.C. Pedersen, B.W. Dickson, and J. Chakareski, "Telemedicine applications of mobile ultrasound," in *Proc. of MMSP '09*, pp. 1–6, 5–7 Oct. 2009 DOI: 10.1109/MMSP.2009.5293344. 107

[224] C. Hewage, M.G. Martini, and N. Khan, "3D medical video transmission over 4G networks," in *Proc. 4th Int. Symp. on Appl. Sci. in Biomedical and Commun. Technologies*, Barcelona, Spain, Oct. 2011. DOI: 10.1145/2093698.2093878. 108

[225] A. Alesanco, C. Hernandez, A. Portoles, L. Ramos, C. Aured, M. Garcıa, P. Serrano, and J. Garcıa1, "A clinical distortion index for compressed echocardiogram evaluation: recommendations for Xvid codec," *Physiological Measurement*, vol. 30, no. 5, pp. 429–440, 2009. DOI: 10.1088/0967-3334/30/5/001. 108

[226] Z. Wang, L. Lu, and A. C. Bovik, "Video quality assessment based on structural distortion measurement," *Signal Process.: Image Commun.*, vol. 19, no. 2, pp. 121–132, Feb. 2004. DOI: 10.1016/S0923-5965(03)00076-6. 108

[227] Metrix_mux objective video quality assessment software, Available: http://foulard.ec e.cornell.edu/gaubatz/metrix_mux/. 108

[228] K. Seshadrinathan, R. Soundararajan, A.C. Bovik, and L.K. Cormack, "Study of subjective and objective quality assessment of video," *IEEE Trans. Image Process.*, vol. 19, no. 6, pp. 1427–1441, Jun. 2010. DOI: 10.1109/TIP.2010.2042111. 109

[229] G. Bjøntegaard, "Improvements of the BD-PSNR model," ITU-T SG16 Q.6 Document, VCEG-AI11, Berlin, Germany, July 2008. 109

[230] E. Kyriacou, M.S. Pattichis, C.S. Pattichis, A. Mavrommatis, C.I. Christodoulou, S. Kakkos, and A. Nicolaides, "Classification of atherosclerotic carotid plaques using morphological analysis on ultrasound images," *Applied Intelligence*, vol. 30, no. 1, pp. 3–23, Feb. 2009. DOI: 10.1007/s10489-007-0072-0.

[231] H. Nasrabadi, M.S. Pattichis, P. Fisher, A.N. Nicolaides, M. Griffin, G.C. Makris, E. Kyriacou, and C.S. Pattichis, "Measurement of motion of carotid bifurcation plaques," *Bioinformatics & Bioengineering (BIBE), 2012 IEEE 12th International Conference*, pp. 506–511, 11–13 Nov. 2012. DOI: 10.1109/BIBE.2012.6399729. 109

[232] C.P. Loizou, C.S. Pattichis, M. Pantziaris, and A. Nicolaides, "An integrated system for the segmentation of atherosclerotic carotid plaque," *IEEE Trans. on Inform. Techn. in Biomedicine*, vol. 11, no. 5, pp. 661–667, Nov. 2007. DOI: 10.1109/TITB.2006.890019.

[233] A. Panayides, I. Eleftheriou, and M. Pantziaris, "Open-Source Telemedicine Platform for Wireless Medical Video Communication," *International Journal of Telemedicine and Applications*, vol. 2013, Article ID 457491, 12 pages, 2013. doi:10.1155/2013/457491. DOI: 10.1155/2013/457491. 116

[234] M. Rahman, A. Aziz, M. Kukar, M.A.N.U. Rajiv, et al., "An optimized speckle noise reduction filter for ultrasound images using anisotropic diffusion technique," *Int. J. Imaging and Robotics*, vol. 8, no. 2, pp. 55–60, 2012. 122

[235] J.L. Fleiss, J. Cohen, and B.S. Everitt, "Large sample standard errors of kappa and weighted kappa," *Psychological Bulletin*, vol. 72, no. 5, pp. 323–327, 1969. DOI: 10.1037/h0028106. 34

[236] G.H. Rosenfield and K. Fitzpatrick-Lins, "A coefficient of agreement as a measure of thematic classification accuracy," *Photogrammetric Engineering & Remote Sensing*, vol. 52, no. 2, pp. 223–227, 1986. 34

[237] S. Golemati, J.S. Stoitsis, A. Gastounioti, A.C. Dimopoulos, V. Koropouli, and K.S. Nikita, "Comparison of block matching and differential methods for motion analysis of the carotid artery wall from ultrasound images," *IEEE Trans. Inf. Techn. Biomed.*, vol. 16, no. 5, pp. 852–858, 2012. DOI: 10.1109/TITB.2012.2193411. 36

Authors' Biographies

CHRISTOS P. LOIZOU

Christos P. Loizou was born in Cyprus on October 23, 1962, received his B.Sc. degree in electrical engineering, a Dipl-Ing (M.Sc.) degree in computer science and telecommunications from the University of Kaisserslautern, Kaisserslautern, Germany, and his Ph.D. degree on ultrasound image analysis of the carotid artery from the Department of Computer Science, Kingston University, London, UK, in 1986, 1990, and 2005, respectively. From 1996–2000, he was a lecturer in the Department of Computer Science, Higher Technical Institute, Nicosia, Cyprus. Since 2000, he has been at the Department of Computer Science, Intercollege, Cyprus and is now a campus program coordinator. Since 2005, he has also been an Adjunct Professor of Medical Image and video processing, in the Department of Electrical Engineering, and Computer Engineering and Informatics, Cyprus University of Technology, Cyprus. He has also been an Associated Researcher at the Institute of Neurology and Genetics, Nicosia, Cyprus since 2000. Dr. Loizou was a supervisor of a number of Ph.D. and B.Sc. students in the area of computer image analysis and telemedicine. He was involved in the research activity of several scientific Cypriot and European research projects and has authored or co-authored 28 referred journals, 54 conference papers, 3 books, and 10 book chapters in the fields of image and video analysis. His current research interests include medical imaging, signal, image and video processing, motion and video analysis, pattern recognition, and biosignal analysis in ultrasound, magnetic resonance imaging, and computer applications in medicine. He is a Senior Member of the IEEE, serves as a reviewer in many IEEE Transactions and other journals and is a chair and co-chair at many IEEE conferences. He lives in Limassol, Cyprus, with his wife and children, a boy and a girl.

CONSTANTINOS S. PATTICHIS

Constantinos S. Pattichis was born in Cyprus on January 30, 1959 and received his diploma as technician engineer from the Higher Technical Institute in Cyprus in 1979, a B.Sc. in electrical engineering from the University of New Brunswick, Canada, in 1983, the M.Sc. in biomedical engineering from the University of Texas at Austin, USA, in 1984, an M.Sc. in neurology from the University of Newcastle Upon Tyne, UK, in 1991, and his Ph.D. in electronic engineering from the University of London, UK, in 1992. He is currently a Professor with the Department of Computer Science of the University of Cyprus. His research interests include ehealth and mhealth, medical imaging, biosignal analysis, life sciences informatics, and intelligent systems. He has been involved in numerous projects in these areas funded by EU, the National Research Foundation of Cyprus, the INTERREG and other bodies, like the FI-STAR, GRANATUM, LINKED2SAFETY, MEDUCATOR, LONG LASTING MEMO-RIES, INTRAMEDNET, INTERMED, FUTURE HEALTH, AMBULANCE, EMER-GENCY, ACSRS, TELEGYN, HEALTHNET, IASIS, IPPOKRATIS, and others with a total funding managed of more than 6 million Euros. He has published 90 refereed journal and 200 conference papers, and 27 chapters in books in these areas. He is Co-Editor of the books *M-Health: Emerging Mobile Health Systems*, and *Ultrasound and Carotid Bifurcation Atherosclerosis* published by Springer in 2006 and 2012, respectively. He is co-author of the book *Despeckle Filtering Algorithms and Software for Ultrasound Imaging*, published by Morgan & Claypool Publishers in 2008 and the revised second edition to be published in 2015. He was Guest Co-Editor of the Special Issues on *Atherosclerotic Cardiovascular Health Informatics*, *Emerging Health Telematics Applications in Europe*, *Emerging Technologies in Biomedicine*, *Computational Intelligence in Medical Systems*, and *Citizen Centered e-Health Systems in a Global Health-care Environment* of the IEEE Transactions on Information Technology in Biomedicine. He is General Chairman of the forthcoming *Medical and Biological Engineering and Computing Conference (MEDICON'2016)* and the *IEEE Region 8 Mediterranean Conference on Information Technology and Electrotechnology (MELECON'2016)*. He was General Co-Chairman of the *IEEE 12th International Conference on BioInformatics and BioEngineering (BIBE2012)*, *IEEE Information Technology in Biomedicine (ITAB09)*, *MEDICON'98*, *MELECON'2000*, and Program Co-Chair of *ITAB06* and the *4th International Symposium on Communications, Control and Signal Processing (ISCCSP 20010)*. Moreover, he serves as an Associate Editor of the *IEEE Journal of Biomedical and Health Informatics*, on the Editorial Board of the *Journal of Biomedical Signal Processing and Control*, and as member of the IEEE EMBS Technical Committee on Information Technology for Health (since 2011). He served as Distinguished Lecturer of the IEEE EMBS (2013–2014), and an Associate Editor of the *IEEE Transactions on Information Technology in Biomedicine* (2000–2012) and the *IEEE Transactions on Neural* (2005–2007). He served as Chairperson of the Cyprus Association of Medical Physics and Biomedical Engineering (1996–1998), and the IEEE Cyprus Section (1998-2000). He is a Fellow of IET and a Senior Member of IEEE.

Printed in the United States
by Baker & Taylor Publisher Services